# Blowout

By the same author
Around and About Ravenglass
Saddle Tramp in the Lake District
Saddle Tramp in the Highlands
Over the Fells
Saddle Tramp on the Isle of Man
The Best Guide to Cumbrian Shows
The History of Harrison and Hetherington Ltd
Around and About Ennerdale
Lakeland Monuments

# Blowout

Robert Orrell

Seafarer Books
2000

Sheridan House
2000

© Robert Orrell 1989, 2000

First published 1989
Robert Hale Limited, London

This edition published in 2000
In the UK by Seafarer Books
102 Redwald Road, Rendlesham,
Woodbridge, Suffolk, IP12 2TE

In the USA by Sheridan House
145 Palisade Street, Dobbs Ferry, N.Y. 10522

UK   ISBN 085036 4 892
USA ISBN 1-57409-109-3

*British Library Cataloguing in Publication Data*
Orrell, Robert, 1934 –
Blowout
1.   North Sea. Offshore structures
Offshore natural gas industries
Offshore petroleum industries
Offshore natural gas and petroleum industries
Personnel. Working Life.
1.   Title
2.   305.9'622

A CIP Catalog record for this book is available from the
Library of Congress, Washington, DC.

Typesetting by Julie Rainford
Cover design by Louis Mackay

Cover photograph shows the Hewett "A" by courtesy of
Brown & Root, Houston, Texas

Printed in Finland by WSOY

# CONTENTS

# ACKNOWLEDGEMENTS

The author is very grateful to the following for their assistance:
Anglia Television; Lord Avebury; BBC Radio Humberside; BBC Radio Merseyside; British Petroleum; Curly Godeaux, Brown & Root, Houston, Texas; Captain Bob Balls MBE; Bristow Helicopters; Commanding Officer, RAF Coltishall; Jacqui Capstick; Department of Transport; Flight Lieutenant Frank Davies; Eastern Daily Press; Warren Elsby, Ron Farrow; Gas Council; Great Yarmouth Library; Institute of Petroleum; Eddie Linguard; Lloyds Maritime Information Services; Meteorological Office; Ivor Nicholas, Barry Rae, Roustabout Publications; Captain Lionel Tomlinson; Dr. John Veitch; Tom Waghorn; Margaret Vincent; Jean Thompson.

This is a true story based on real people and actual events, but with a few exceptions the names of people have been changed. The following appear under their real names: Paul 'Red' Adair, Vernon Andrews, Capt. Bob Balls, Connie, Richard Craig, Frank Davies, Len Dawson, Dick Drake, Ronnie Farrow, Ken Forsyth, Gerry Freeze, Harry Goodison, Ron Gower, Tommy Grimes, Larry Hoelscher, Bill McMillan, John Pottage, Gordon Ray, Joe Robinson, Revd. George and Mary Shelley, Louis Sylvester, Lionel Tomlinson, Al Vick and J.C. Youngblood.

# Illustrations

Picture Credits
Robert Orrell, 1-6, 8-9, back page; Al Vick, 7; Syndication International, 10; The Press Association, 11; Assignments Photographers / Yarmouth Press Agency, 12.

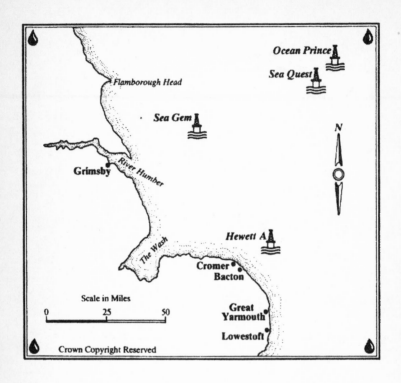

Ocean Prince

Sea Quest

Flamborough Head

Sea Gem

*N*

*River Humber*

Grimsby

Hewett A

*The Wash*

Cromer
Bacton

Scale in Miles

0          25          50

Great
Yarmouth

Lowestoft

# 1     HIT THE TRUCK, RADIO MAN

A freezing fog hung over Great Yarmouth harbour, and frost glistened like silver on the silent ships as I drove slowly along the cobbled dockside. Somewhere in the greyness a clock started to chime, and I glanced at my watch. It was five o'clock on a bitterly cold morning in November, and I was so tired I could scarcely keep my eyes open.

The orange glare of roadside lamps guided me like beacons through the patchy fog as I threaded my way through a maze of derelict warehouses and the boarded-up offices of trawler companies, fish salesmen and ship's chandlers. Here and there bulldozers had demolished the old buildings, and occupying new offices and storesheds surrounded by wire-fenced compounds were companies with strange-sounding names: Amoco, Conoco, Odeco, Halliburton, Dowell Schlumberger, Magcobar, Baker Oil Tools.

Two patrolmen in a parked police car eyed me suspiciously when I shone a torch on the name-plate of one of the new office-blocks, and when they waved at me to stop, I pulled alongside and wound my window down.

'Can you tell me where the Phillips Petroleum office is?' I said.

Ignoring the question, the driver climbed out of the patrol car, pulled my car door open and told me to get out. 'And if you want to argue about it,' said his mate, a sergeant, gripping my arm menacingly, 'I'll have to help you.' I got out.

The cold air cut through me, and I shivered on the pavement while they checked my number-plate, lifted the seats in the rear of the car, emptied my pockets, then radioed the details of my driver's licence to their head-quarters.

'Bit early to be calling on an oil company,' said the sergeant sarcastically. 'What do you want 'em for?'

'I've got a job on a rig,' I replied, 'and I've to be at their office by eight o'clock.' I fumbled in my coat pocket. 'Here's a letter that'll prove it.'

The sergeant snatched it out of my hand and studied it by the light of his torch. 'It looks genuine enough,' he grunted, handing the letter back. 'Keep going up this road; the office is on the right. A low building with the big sign, "Phillips 66", on the wall. Be careful you don't miss it in the fog.' The patrolmen climbed into their car and were about to drive away when the sergeant wound his window down and said, 'If you turn left at the end of the next block, there's a café the dockers use. Bit of a dump, but at least you'll get a cup of coffee.' The driver accelerated and the car was swallowed up in the fog.

The café was a dingy shop wedged between a newsagent's and what in the prosperous days of the fishing boom had been a net factory. The stench of rancid fat and human sweat hit me when I opened the door, and I stood on the threshold undecided whether my stomach could take it at that hour of the morning. The room was empty except for a large, unshaven man seated behind the counter, a cigarette hanging from a corner of his mouth, engrossed in a girlie magazine propped against a plateful of sandwiches. He eyed me up and down angrily, then spat the cigarette onto the floor.

'Either come in or piss off out, boy,' he snarled. 'You're lettin' the fog in, and it's cold enough to freeze yer balls off.'

'Sorry,' I said, and closing the door behind me I sat down at a table.

'It's self-service,' he snapped. 'There's no waiters here!'

I went to the counter and studied the 'menu' scrawled in chalk on a blackboard.

'You a cop?' he quizzed, staring at an ID card in my wallet when I opened it to take some money out.

'No,' I replied. 'I'm going out to a rig. I'll have a mug of coffee and a cheese sandwich.'

He shoved the girlie magazine to one side, lifted a sandwich off the plate and slapped it down on the counter. Reaching for a mug, he filled it from a coffee-machine and placed it by the sandwich.

'If you're looking for a woman, I can give you a phone number,' he said bluntly.

I laughed and gulped a mouthful of coffee. 'I've been driving all night,' I said, 'and I'm that tired I can hardly raise my head, let alone anything else.'

He thought it was a huge joke, and the rolls of fat on his large belly shook as he bellowed with laughter.

'Hey, what do yer think of her?' he said suddenly, thrusting the girlie magazine at me.

In the photograph a hard-faced naked blonde with blotchy skin lay on a bed with her legs wide open. Before I could say anything, he grabbed the magazine back and leered at it.

'Jeez, I could do something for her.' Holding the photograph against his crotch, he jerked against it, then placed the magazine back against the pile of sandwiches.

'The Dutch lads bring 'em in off the ships,' he confided. 'Get you anything you want. Photo mags, blue movies, cigars, watches; you name it, and they'll get it.'

I ordered another coffee and pulled the drilling company's letter out of my pocket and read through it. It was terse and to the point. 'We confirm your appointment as radio-operator on the Hewett "A" drilling-platform. You are to report to Phillips Petroleum Company, South Denes Road, Great Yarmouth, on November 8, 1967, at 0800 hrs, ready to sail on the MV *Inkenturm* at 0900 hrs.'

I had no idea what I was letting myself in for. The nearest I had ever been to an oil rig was when I was at sea in the early 1960s and we passed one under tow in the Bay of Biscay. But fate plays strange tricks. I had been the skipper of a yacht belonging to an oil tycoon, and soon learnt that the mega-rich live in a protected artificial world that is planets away from the lives of real people; so the owner and I parted company. I had been out of work for some months when to my surprise he phoned to say that a North Sea drilling consortium he was involved in was looking for a qualified radio-operator to work on a new rig. If I was interested, I should call KCA Drilling in London immediately. Jobs at sea were hard to find at the time, and anything was better than trying to survive on unemployment benefit with a young family to provide for, so I accepted the job and was assigned to work with Rimrock Tidelands, a Louisiana drilling company contracted to drill for Phillips Petroleum.

The café door opened and two tough-looking men wearing blue reefer jackets pushed their way in and kicked the door shut behind them.

'Mornin', Billy. Mornin', Charlie,' greeted the café-owner as he filled two pint mugs with tea. 'You bin called out early then?'

'Yeah,' replied one of them as they slumped into chairs at an empty table. 'A Smit Lloyd boat had to take some urgent stuff out to one of the rigs.'

The café-owner carried the mugs of tea over to the table, sat down and handed out cigarettes. The conversation dropped to a murmur. I drained the last of my coffee, left the stale cheese sandwich untouched and went out into the freezing air.

The fog had lifted, and above the street lamps a thin grey line of dawn was beginning to break through the black sky. A cargo ship, ablaze with lights, was steaming slowly down the harbour, and in the stillness I could hear the dull thud, thud, thud of its engines and the shrill ring of the telegraph when the captain ordered more speed. With plenty of time in hand, I was in no hurry, and for half a mile I kept pace with the captain and his ship until we parted on our separate adventures, he heading out into the open sea and I driving into the car-park of Phillips Petroleum Company. The office lights were on, and I opened the door and walked into a carpeted reception area furnished with easy chairs and low tables, each stacked with a neat pile of oil-industry magazines. In one of the chairs a heavily built man with a large cigar jammed between his teeth was stretched out, the heels of his tooled cowboy boots resting on one of the tables. He looked up from the newspaper he was reading and stared at me with unflinching blue eyes.

'Hi,' he said, still gripping the cigar in his teeth. 'What can I do for ya?'

'I was told to report here,' I said. 'My name's Bob Orrell – I'm the new radio-operator for the Hewett "A".'

I handed him my letter of appointment and he studied it for a minute or two through a haze of cigar smoke, then held out his hand.

'Glad to know you, Bob,' he said, shaking my hand warmly. 'I'm Al Vick, the rig superintendent.' Then laughingly he added, 'You ain't the noo radio-operator, you're the only goddam radio-operator; the Hewett ain't finished buildin' yet!' He stubbed his cigar out in an ashtray, swung his feet off the table and stood up. 'Wanna cup of coffee?' Without waiting for me to reply, he strode into an adjoining office and re-appeared almost immediately balancing a carton of milk, a bag of sugar, two paper cups and a jug of coffee in one hand, and clutching a paper bag in the other. Lowering everything safely onto a table, he poured the coffee.

'There you go,' he said, handing me one of the cups, 'and if ya feeling hungry, there's a coupla fresh doughnuts in the bag.'

I chewed a doughnut and washed it down with coffee while Al talked about the problems oil companies were having drilling for gas and oil in the wild North Sea.

'There ain't no weather like it in the whole goddam world!' Al exclaimed, putting a match to the end of a fresh cigar. He inhaled deeply and blew a cloud of smoke to the ceiling. 'Just when ya get a rig ready to start drillin', a gale gets up and blows the sonofabitch off location – it knocks hell out of schedules. When we're drillin' off Texas, at least we know when the hurricane season is, but in the UK the only way you can tell it's summer is when the goddam rain gets warmer.'

A phone rang in an office, and Al went to answer it. I filled my cup with coffee and browsed through a magazine lying on a table. In it there were comments from the executives of American oil companies, speculating on the future of world exploration for oil. Most of them expressed only a cautious optimism about the fortunes that might be made in the North Sea, but the ruthless expressions on the hard-bitten faces in the photographs showed that they were each determined that at the slightest whiff of oil it would be their company that would beat the others to the jackpot. Pages of photographs showed many different types of rigs and platforms, and they all seemed remarkably flimsy constructions. I lost interest in my coffee and doughnut when, at the foot of a page outlined in black, I read an account of the loss of the rig Sea Gem which had collapsed in the North Sea off the Lincolnshire coast in 1965 with the loss of thirteen men, including the radio-operator!

Al returned with a fresh jug of coffee and the news that the weather forecast was for a bright, sunny day and a calm sea, and said that the boat going out to the Hewett would be leaving from a quay a few miles away on the east side of the harbour.

'I've got a few chores to do,' he said, 'so relax and help yourself to coffee. There's a whole bunch of guys goin' out to the rig, and I'll get someone to give you a call when the transportation arrives.'

I picked up the newspaper Al had left behind, but it was full of depressing accounts of the war in Vietnam and I dropped it back on the table. Office staff began to trickle in, and by eight o'clock it was bedlam, with phones ringing in every office, telex machines chattering non-stop and people rushing about the building as if the world was about to end. A minibus pulled up outside, and when the driver hooted, four men, carrying briefcases and dressed in high-heeled boots, carefully pressed trousers and a variety of colourful fur-lined coats, all wearing wide stetson hats, came out of an office and trooped through the reception area. As the last one reached the door, he stopped and looked back.

'You the radio man?' he drawled. I nodded and said yes. He jerked a thumb towards the minibus. 'OK, feller, hit the truck.'

At the quayside the German supply boat *Inkenturm* was loaded and ready to leave, and as soon as we stepped on board, the mooring-ropes were cast off. Plumes of black smoke poured from the twin funnels as the big diesel engines roared into life, and we steamed briskly down the River Yare and into the North Sea. It was a beautiful morning, still bitterly cold after the frost, but clear and sunny with a calm sea and a blue sky. There were a number of other passengers on the boat, and everyone went below to crowd into the warmth of the crew's mess-room, but it was too nice a day to miss, so, zipping up my windproof coat and pulling a woollen balaclava over my ears, I stood on deck and enjoyed the movement as the ship gently rolled and dipped in a slight swell. The *Inkenturm* was one of a new breed of ships developed by the Americans as a workhorse to meet the needs of the offshore drilling industry. Driven by two powerful engines and bristling with modern gadgetry and every conceivable aid to navigation, it made the typical British coasters I had sailed in look like relics from a steamboat museum.

We had been at sea for about an hour, and I was busy taking photographs, when one of the Americans who had travelled in the minibus appeared on deck, fanning himself with his stetson hat and mopping his brow with a large handkerchief.

'Jesus H. Christ!' he exploded. 'It's hotter than a mule's ass down there, and that Kraut steward says there ain't no beer on board. Lying sonofabitch.' Flinging his briefcase onto the deck, he sat on it and lit a cigar. 'You ain't no Phillips man,' he said, staring at me coldly. 'What's your company?'

When I told him that I was a radio-operator and worked for a British drilling company, he burst out laughing and sneered, 'A Limey drillin' company! Boy, you must be kidding! The Brits couldn't drill through shit with a laser beam. When I was in Iran, your BP company spent more time fillin' in a bunch of forms than they did drillin'.'

His blustering arrogance made me really angry, but before I could say anything, he was called back down below to join a card school.

'You should have told big-mouth that BP struck gas in the North Sea years before Phillips,' said an English voice. I turned to find an athletic-looking man in his thirties grinning at me from a doorway. 'Don't let smart-arse upset you,' he said. 'Not all Yanks are like him. I hear this is your first rig.'

He laughed when I looked surprised. 'How the hell do you know that?' I asked.

'Well, people talk in this business, and it's amazing what you learn on the grapevine. I don't work for Phillips, by the way. I'm an engineer with Brown & Root, the company that's building the Hewett platform. The accommodation packages are not in position yet, so you'll be living on the *Hercules*, our crane barge, until it's finished.' He opened his briefcase and pulled out a small book. 'Read that,' he said with a laugh. 'It'll help pass the time.'

Wedging myself into a corner of the deck, I read through the book, and it was hilariously funny. Produced by the American Embassy in London with the object of advising Americans about the British way of life, it contained gems of wisdom like, 'Everything here is very, very British. It is not very, very English and you should learn the difference ... Don't be boastful about how your country does things ... The stranger to England might never understand the history of the local pub, much less the hours it keeps and the reasons behind the existence of both.'

When I handed the book back, I was shaking with laughter.

'I thought you'd enjoy it,' he said with a big grin. 'I like the one about don't be boastful about how your country does things. Telling you how everything is much better in the States is a Yank's favourite pastime.'

By mid-morning several specks had appeared on the horizon, and, as we drew nearer, a buzz of interest went through the passengers. The card school was temporarily abandoned while everyone lined the ship's side to stare through binoculars or take photographs.

'Have a look at your new home,' said my engineer friend, handing me his binoculars. 'That's the platform on the left of the crane barge.'

At first only a blurred mass filled the lens but, as I turned the focusing-knob, there, protruding from the blue sea, was a flat table-top supported on eight green-painted tubular legs. It was incredibly exciting, and I stared at it for a long time, aware that I was witnessing history in the making. A flotilla of tugs, fussing around the enormous hulk of the crane barge *Hercules*, was dwarfed by the jib of the gigantic crane that towered above the barge and filled the sky.

'Look to the right of the barge,' instructed my friend. 'You'll see a tug towing a barge, with what looks like big grey boxes on it.'

'Got it,' I said, swinging the binoculars round.

'Those are the accommodation packages and the drilling works,' he went on. 'They've been built in Holland and weigh about 400 tons apiece. If the weather holds, the crane will lift them onto the platform in the next few days.'

The tugs had finished anchoring the crane barge by the time the *Inkenturm* slid neatly alongside, and immediately a small crane swung into action lifting cargo off the deck. Long, flexible pipes were coupled to hydrants on the ship's deck and, at a signal, pumps in the engine-room blasted thousands of gallons of diesel oil and fresh water into tanks deep inside the barge. Ever aware that the North Sea weather could change in a matter of minutes, the *Inkenturm*'s captain paced impatiently in and out of the wheelhouse, bawling orders in German. There was no time to rig a gangplank for the benefit of passengers. We threw our bags and cases onto the *Hercules* and clambered after them.

Unlike the Texas oilfields of the USA and the desert oilfields of the Middle East, where the possibility of striking oil or gas was almost guaranteed, the results of the early searches for oil and gas in Europe were so disappointing that most companies considered it a waste of money. In the 1930s British Petroleum, having drilled exploratory holes offshore in the English Channel without success and then later discovered a small gas field in Yorkshire, concentrated the bulk of their exploration overseas. It was the persistence and optimism of a partnership between Shell and Esso, convinced that gas or oil was to be found on the European mainland, that finally paid off and led to the rush to explore the bed of the North Sea.

Near the quiet village of Slochteren in the north of the Netherlands, farmers and their families were busy working in the neat, square fields that spread like a patchwork quilt across the flat Dutch countryside. It was August 1959, and the corn was ripening to a rich golden colour ready for harvesting the following month.

In one of the fields some strangers had arrived and built a tall steel tower, and day and night it vibrated with the roar of machinery as a long metal pipe slowly bored a hole deep into the ground. At the foot of the tower a man in mud-stained clothes gripped a lever in his gloved hand and stared at an indicator on a large dial. It twitched, then twitched again, then again! The strangers threw their hats in the air and went wild with delight. The farmers and their families looked up for a moment, wondering what all the fuss was about, then got on with their work. How were they to know that the strangers had tapped an underground reservoir of gas that was to affect the lives of everyone in the country?

The Slochteren gas field was one of the largest ever discovered, and the governments of the countries bordering the North Sea cast envious glances in the direction of the tiny Dutch vil-

lage. Geologists in Britain, France, Germany and the Scandina-
vian countries stampeded to get out their rock-formation maps to
study the structure of Western Europe. If there was gas under the
ground in the Netherlands, there must be gas under the ground in
the countries close to them - but where? The geologists knew
that millions of years ago the British Isles, the North Sea, Scan-
dinavia and Europe were one vast mountainous region. In later
times, due to movements of the earth's crust, seas invaded the
land, and as plant and marine life died it sank to the bottom and
decayed. As the earth moved again, the decaying matter was
compressed and formed the coal deposits of Britain, France,
Germany and Belgium. Later still, a further upheaval caused a
sea to cover Western Europe which later evaporated when the air
became hot and arid, leaving behind a deposit of salt thousands
of feet thick, conveniently covering the coal deposits, which
through pressure and heat were now giving off large quantities
of gas. In 1937 the western extremity of this gas concentration
was discovered and tapped near Whitby in Yorkshire, by a company which later became British Petroleum, but the production
rate was low. Twenty years on, the partnership between Shell
and Esso struck gas on the eastern extremity of the gas field with
their huge find at Slochteren. The geologists' conclusion was
simple: if there was gas at the western and eastern edges, there
must be gas somewhere in the middle. There was just one snag.
A few thousand years ago the earth had moved again and the bit
in the middle was flooded. Map-makers called it the North Sea.

When it was realized that there was a strong possibility of
finding gas under the North Sea, the inevitable arguments arose
between countries bordering the sea about who owned the sea-
bed. Years of wrangling and political heel-dragging went by
before an agreement was finally ratified in 1964, and on Boxing
Day the same year the American jackup rig Mr Cap entered the
history books by drilling the first well in the British sector for
American Overseas Petroleum. The rig also drilled a well for the
Shell company, but both results were disappointing and, having
had a taste of North Sea weather with hurricane-force winds, it
pulled up its legs and departed for the warmer climes of Nigeria.
The supply boat attending the rig at the time was a converted

trawler called the *Hector Gannet,* which had been through many adventures and which, years later, was to feature so tragically in the evacuation of the Hewett "A".

After Mr Cap's departure, the drillship *Glomar IV* was brought from Texas to drill for the Gulf Oil Corporation, but it dragged its anchors in a fierce storm and broke the drill pipe. Staying only long enough to pull the rest of the pipe out of the hole, *Glomar IV* weighed anchor and followed in Mr Cap's wake to the sun.

For those able to put up high stakes in the chancy game of North Sea roulette, the wheel of fortune was still spinning, and when, in 1965, the ball came to rest, it was on the name of the UK's leading oil company, British Petroleum. That year saw the first major success in the search for gas and also the first drilling-rig disaster.

British Petroleum had placed an order with the Belfast shipyard of Harland & Wolff for a new drilling rig, but in the meantime they were keen to test a location off the Lincolnshire coast which geologists believed could yield gas. A drilling barge, the Sea Gem, was chartered from a French company and towed to Middlesbrough for the fitting of accommodation packages and drilling machinery. One day at the beginning of June 1965 the Sea Gem was towed out to its location, about forty miles east of the Humber, and soon after the ten gigantic legs had been jacked down, drilling started. The laborious work went on all through the summer, and by September the drill had reached nearly 9,000 feet, but there was no gas. All the signs seemed to indicate that the well was dry. But the drill bit was kept rotating, and on Friday 17 September perhaps the most momentous event in the saga of North Sea exploration occurred. Traces of gas began to register on the rig's instruments, and as drilling continued, it soon became obvious that it was more than just an isolated pocket. It was the first major discovery of natural gas, hailed as 'the North Sea Klondike', but the price paid for it was in human lives.

BP decided to continue drilling the well to a depth of 10,000 feet, then move the Sea Gem to another location about two miles away. By Christmas the well had been plugged and the rig was made ready to move.

Just before midday on Boxing Day, Ronnie Farrow, the head chef, tossed and turned in his bunk, unable to sleep. He had been working on night shift and was desperately tired, but the screech of metal upon metal as the rig was being jacked down towards the sea made sleep impossible. Suddenly the rig lurched violently to one side, and he was thrown across the cabin onto the bunk of his cabin-mate. Heaving open the cabin door, both men, clad only in their underpants, ran to see what was happening. The rig heeled over alarmingly and pans, food and crockery were strewn all over the galley floor. Sensing that the rig was going over, Ronnie ran back to his cabin, grabbed a coat and dashed up on deck to where he knew there was a lifeboat by the radio-operator's shack, but it had gone. To his amazement, he saw the radio shack floating on the sea, and the lifeboat alongside with the radio-operator sitting in it. Ronnie yelled at him to bring the boat to the rig, but he seemed dazed and just sat staring into space.

By now the rig was listing at a steep angle and, desperate to reach the boat, Ronnie jumped into the sea, only to be lifted back again by a big wave which swept him under a metal companionway, trapping his legs under a heavy object. As the rig tilted further, the water rose above his head. He had almost given up hope when Ken Forsyth, one of the drilling crew, grabbed him by the arm and pulled him to safety. By a stroke of fortune, the object that had trapped his legs was a life-raft, and they quickly inflated it and jumped in. More men clambered aboard until it was packed with wet, shivering bodies. They were almost clear of the rig when, with a roar, it turned right over, and they looked up horrified to see one of the gigantic legs falling towards them. It missed the life-raft by only a few feet, and they managed to paddle to safety. The last Ronnie saw of the unfortunate radio-operator, he was still sitting rigid in the centre of the lifeboat as it drifted away.

Of the thirty-two men on board, thirteen were lost, including five of Ronnie's catering staff and the radio-operator. The official accident report concluded that the Sea Gem had collapsed due to metal failure in the jacking system.

By an incredible stroke of luck, the crew of a passing
steamer, the *Baltrover,* saw the rig fall and went to the rescue. At
the Bristow helicopter base at Tetney, near Grimsby, the scheduled crew-change helicopter took off for the Sea Gem, completely unaware of the disaster. The pilot, Captain Bob Balls, set
his navigation instruments, then headed out over the sea towards
the rig. He was puzzled that there was no reply when he called
the rig on the radio, but presumed that the radio-operator was
busy, and flew on. Visibility was excellent but when, only five
miles away from the rig, he could not see it or get a response to
his radio calls, he sensed that something was wrong. Arriving at
the rig position, he saw a ship picking up men out of life-rafts
and, flying low over the sea, he counted a number of bodies
floating on the surface. But there was little he could do. Passen-
ger-carrying helicopters were not equipped with lifting-winches.
He dropped a life-raft into the sea, hoping that someone would
be able to reach it, then reluctantly flew back to Grimsby.

Coming so soon after BP's remarkable triumph, the Sea Gem
tragedy was a sad irony, but in an industry that pits itself against
the forces of nature the lessons learnt are often the result of bitter
and tragic experiences.

Though the North Sea's first major rig disaster rocked the industry, it did little to dampen the determination of the twenty or
so groups of companies which had bought drilling concessions
in the British sector, and the demand for rigs far exceeded the
number available.

As with most things concerned with the oil industry, it was
the Americans who first built rigs capable of drilling offshore. In
the late 1940s, when it was discovered that the Louisiana oilfield
extended under the Gulf of Mexico, the resourceful oilmen developed a flat barge fitted with movable legs and a drilling derrick. The concept was simple. The barge was towed out to a location, and the legs were lowered onto the sea-bed, then the
barge jacked up the legs until it was clear of the sea. The drillers
then had a stable platform to work from. When the well was
completed, the barge was jacked down to float on the water, the
legs were lifted off the sea-bed, and the rig was towed to the next
site. The ill-fated Sea Gem was this type of rig, and it had the

disadvantage of being restricted by the length of its legs to drilling in comparatively shallow water. Jackup rigs were seldom able to drill in depths of more than 200 feet.

To enable drilling to be carried out in deeper water, a semi-submersible was developed. Basically it was a flat platform supported by fixed vertical legs welded to large tanks or pontoons. The pontoons provided buoyancy so that the rig floated, or when flooded with water they acted as ballast to lower the platform in the water, providing stability for drilling. On location the rig was held in place by several large anchors. To operate round the clock far out to sea, both the jackup rigs and the semi-submersibles were provided with crew accommodation and storage facilities for essential equipment.

Another device used for drilling in deep water was the drill-ship. In effect it was a cargo ship fitted with a derrick and converted so that drilling could take place through a hole in the centre of the hull. Unlike barges and semi-submersibles, which could be moved only with the aid of tugs, the drillship had its own propulsion power. Some drillships were held on location by anchors, but as technology improved, a system of dynamic positioning was introduced which automatically controlled a series of propellers.

American drilling contractors wanted new rigs in a hurry, and the orders to build them injected new life into hard-pressed British industry. In the north-east, Clydebank and Belfast, strange constructions began to take shape on the slipways that had been the birthplace of many famous ships. Drawing on American design expertise, British-built rigs were soon hard at work in the search for the gas bonanza. The media avidly followed the fate and fortunes of this new and romantic breed of seagoing hardware, and names like North Star, Constellation, Orion, Norsmec I, Ocean Prince, Orient Explorer, Sea Quest, Staflo and Transocean II became as familiar to the British public as the *Queen Mary* and the *Ark Royal*.

The financial costs involved in exploring the North Sea were staggering, and the competition between oil companies was fierce. Understandably, they kept information about their progress a closely guarded secret, but when they did strike lucky, they liked the world and their competitors to know about it. Early in February 1966 the headline in the Great Yarmouth local paper, the *Yarmouth Mercury,* proclaiming 'North Star Strikes It Rich', gave the first news of a find by the International Drilling Company rig working for one of America's major oil companies, Phillips Petroleum. 'It does appear to be very significant,' said Mr Larry Hoelscher, the company's drilling superintendent cautiously, but one of his aides put it in a more succinct American oilman's way: 'It's a whacker!'

The discovery, on the Hewett Bank, twenty-three miles northeast of Great Yarmouth, was acknowledged as having a higher flow rate than any other, and what made it of exceptional value was that the nearest point on the Norfolk coast was only fifteen miles away. The nearer the coast, the less the cost of laying pipe under the sea-bed would cut into potential profit. Profit was the major concern of the oil companies; not surprisingly, they were keen to recoup the high cost of exploration by squeezing every penny they could out of the sale of gas from production wells, and it led to many protracted negotiations.

The British bulldog, in the form of the Gas Council, and the obstinate American buffalo snarled and roared at each other, then retired to their respective lairs to issue statements. Speaking about the Hewett find, a Phillips spokesman declared huffily that his company did not intend to make any plans for the development of the field until the gas-price problems were solved. The Gas Council, equally obdurate, declared that their offer was final. Despite the heated rhetoric, both sides knew very well that gas lying in the ground was of no value to anyone, and eventually an agreement was reached. Some oil companies accused Phillips of a 'sell-out', but presumably the Phillips accountants had done their sums, for they assured shareholders that once the gas started to flow in, so would the dollars.

Anticipating an agreement, Phillips had already made contingency plans, and no time was lost in commissioning the Texas construction company of Brown & Root to fabricate a platform and erect it on the Hewett bank. Rimrock Tidelands of Louisiana were contracted to drill several directional holes to radiate out over the gas field like spokes of a wheel.

All the problems seemed to be over - except one. The drilling contractors had to hire a radio-operator. Phone wires hummed, telexes chattered and telegrams winged their way round the world. The messages were the same. Phillips wanted the best available at the cheapest rate. Once again they had to compromise. They got me!

Seen from deck-level, the crane barge *Hercules* was a mind-boggling clutter of tanks, lengths of pipe, portable compressors, welding-bottles, drums of oil, coils of wire rope and deckhouses bristling with radio antennas. On the stern, the turntable supporting the massive crane occupied almost the full width of the deck, and above it, supported by cables as thick as a man's arm, the steel latticework of the 200-foot jib soared to the sky.

Diesel engines thundered, a public-address system constantly blared out instructions, and men in blue overalls and shiny aluminium safety-hats swarmed like ants over the greasy wet deck, adding to the racket with hammers, drills and metal-cutting saws. In the midst of the confusion a small red-painted portable crane, mounted on rubber caterpillar tracks, trundled backwards and forwards, moving pallets of stores and metal baskets full of pipe fittings, like a busy housewife determined to create some order among the chaos.

It was a completely new world to me, and it was several days before I realized that, far from working in an uncontrolled turmoil, the Texans ran the barge and the work programme with clockwork precision. With the aid of anchors and tugs, the *Hercules* was heaved alongside the skeleton of the platform, power cables were strung across and teams of riggers, welders and electricians worked round the clock making the platform ready for the drilling works and accommodation packages. When the strong tidal flow was at its weakest, I watched as a diver was helped into a thick rubber suit and had a metal helmet fastened over his head before he climbed awkwardly down a ladder to sink beneath the surface in a flurry of bubbles.

While a helper carefully paid out the diver's safety-line and air-supply pipe, the diving boss, a gum-chewing giant in a thick bearskin coat and a Mexican sombrero held down with a plaited leather chinstrap, kept in contact over a two-way intercom.

'You OK down there, Sonny?'

The sound of heavy breathing came from the receiver, and a distorted voice answered, as if someone was trying to gargle and talk at the same time: 'Yeah, I'm on the bed, I guess. Sure is a helluva lot of fish down here.'

For fifteen minutes or so the diver checked one of the platform's legs for corrosion, but as the tide turned, he was unable to stay on his feet and was hauled back to the surface.

Below deck there were offices, stores, a radio-room well equipped with powerful radio-telephones, a comfortable recreation-room, a large galley and dining-area, and accommodation for over seventy men. Americans have long had a reputation for looking after their men, even under the most difficult conditions, and the food was unbelievably luxurious. Steaks as big as a dinner plate were standard fare, and the choice and presentation of the food were equal to that in any fashionable London hotel. In stark contrast to the grease and grime of the working deck, the living-quarters and dining-area were kept spotlessly clean. Unlike the average British factory canteen, where workers in filthy dungarees soil tables and chairs, absolutely no one was allowed in wearing working clothes.

Television was very much in its infancy in the 1960s, and off-duty times were spent reading, playing cards or watching films on an ancient 16mm projector. Whether by accident or choice, the movies were invariably Westerns, and the recreation-room would be crammed with Spanish welders shouting and cheering as a cowboy hero mowed down hordes of Indians, or a Texas Ranger, braving a hail of bullets, saved a weepy heroine from a fate worse than death in the clutches of a swarthy villain with a stubbly chin and bad teeth.

The barge radio-operator was a cheery Yorkshireman called Keith, and in between interruptions when one of several radio sets bellowed for his attention, he gave me an insight into the American attitude to life.

'What I like about the Yanks who work in oil business,' he confided, 'is that they don't worry too much about paper qualifications. If you can do the job they're paying you for, you'll be OK, but make one mistake that costs 'em money and it doesn't matter whether you're the president of the company or the shit-house-cleaner, you're fired!' He turned to a desk to write down a weather forecast being broadcast by a coastal radio station, then went on: 'They like to think the almighty dollar rules the world, but underneath all the bullshit they're a great bunch to work with.'

Apart from occasional short spells standing in for Keith, I had little to do. No one from the drilling company I was to work for or from Phillips Petroleum made themselves known to me, and the friendly engineer I had met on the *Inkenturm* had been trans-ferred to Australia at short notice. I went to the bargemaster's office, explained who I was and asked if anyone had been look-ing for me. He had been up all night dealing with a breakdown, and as he slurped a mug of coffee he squinted at me with red-rimmed eyes.

'Son,' he sighed wearily, 'you can bet your ass if those guys wantya they'll sure as hell holler. If you want somethin' to do, go take your Limey cock on deck and piss over the side.'

Keith could hardly stop laughing when I went back to the ra-dio-room and told him what had happened.

'He's an ignorant old bastard,' I fumed. 'There's no need for that!'

'You're lucky that Curly was so polite,' chuckled Keith, wiping his eyes with his sleeve. 'You must have got him in one of his happier moods!'

Word went round that the drilling works was to be lifted onto the platform the following day, and I was up early with my cam-era to get a good vantage-point. At 200 tons, it was the lightest of the three packages, but the engineers were taking no chances. It was about fifty feet from the deck of the barge to the top of the platform, and every angle was studied carefully before the tugs were signalled to nudge the barge carrying the drill works against the side of the *Hercules*. Wire slings, shackled to the package, were slipped over the giant hook of the crane, and mo-tors whirred. For a long time nothing happened as the tension was taken up on the miles of wire, but then the slings went taut and the package was lifted very slowly into the air. Everyone stopped work and held their breath as the load rose high above the sea. When it was a few feet above the level of the platform, the crane edged round on its turntable and lowered the package neatly into place. I half expected the onlookers to cheer, but to them it was all in a day's work. A lot of the crew had been with Brown & Root for years and had seen it all in the States, the Persian Gulf, Africa and the Far East.

The evening forecast warned of approaching bad weather, and the crew worked feverishly in the hope of lifting the engine-room package, which housed the huge diesel engines that powered the drilling works and generated electricity. The tugs nudged the barge against the *Hercules,* the slings were hooked on and the mammoth grey box began to lift slowly into the air, but then it stopped and went down again. The crane motors whirred, but again the load was lifted only a few feet above the barge, then lowered down. A loudspeaker on deck blared and Curly, the bargemaster, shot up from below deck and scrambled up the metal ladders into the control room of the crane. A few minutes later he reappeared at the foot of the ladder, flung his safety-hat to the deck in a rage, then kicked it into the air. The weather-beaten faces of a group of crew men standing nearby paled, and they quickly made themselves scarce. The burly diving boss looked up from the portable compressor he was working on and spat a wad of gum into the sea.

'Well, waddya know,' he drawled laconically, 'ah do believe Curly's done gone and blown a fuse.'

When the reason for Curly's spirited demonstration emerged, the news was devastating. A fault had been discovered in the crane's wire lifting-rope. It was thousands of feet long, and every inch would have to be replaced before a heavy load could be moved. The great jib was lowered to the deck, and later that afternoon a fierce gale howled in from the west, pounding the *Hercules* with tremendous waves and drenching anyone unfortunate enough to be working on deck. Tugs, called in to pull the *Hercules* away from the platform, plunged and rolled in the heavy seas and at times almost disappeared from view when a particularly violent wave swept them from stem to stern. By midnight the gale had increased to storm force, and a 60 mph wind screamed through the rigging like a thousand demons. Reflected in the deck lights, the seas were awesome. All work on deck was stopped and everyone sent below. I went to my bunk but the rolling motion of the barge kept me awake and, taking an armful of sailing magazines, I wedged myself in an easy chair in the recreation room.

· The nightly film show had been abandoned when the barge lurched into a big wave and catapulted the projector into the audience, and the room was empty except for the diving boss and four or five Americans engrossed in a game of poker. Somehow they managed to keep the cards and money on the table and stay in their seats, and it could easily have been a studio set for a Wild West movie. Apart from his usual Mexican sombrero, the diving boss was decked out in a black embroidered shirt with white tasselled pockets, black pants held up with a wide leather belt which fastened with an ornamental silver buckle, and black high-heeled cowboy boots. The others were dressed in similar if more colourful regalia, preferring loud check shirts and brown tooled boots. Large stetson hats were standard headgear, and the men all gripped large cigars in their teeth. The only items missing were Colt 45s.

As the night wore on, the heap of cigar butts in the ashtrays and the stack of British banknotes in the centre of the table rose higher and higher, and when a player dropped out, his place was taken by someone else. There was no let-up in the ferocity of the gale, and the barge continued to pitch and roll heavily, but the cards were dealt again and again. I watched mesmerized as £5, £10 and £20 notes fluttered onto the table like confetti. Towards dawn the strain began to tell, and the players dropped out until only the diving boss and a thin, weasel-faced man were left. The others gathered round the table, and the tension mounted as the stakes rose, but no one dared speak.

'Nine, ten, jack, queen, king,' grinned the diving boss, triumphantly spreading the cards on the table. No one moved as weasel-face studied his cards. With a shaking hand and breathing heavily, he dropped them onto the heap of money.

'I've got a royal flush!' he croaked.

The onlookers whooped with delight as weasel-face pulled the money towards him and started to count it.

'Well, I'll be doggone!' exclaimed the diving boss, leaning back in his chair and pushing the sombrero to the back of his head. 'Gus, you sonofabitch, I thought I had you there.' He climbed stiffly to his feet. 'I guess I'm cleaned out. C'mon, you guys, let's go get a coffee and a burger.'

They left Gus counting his winnings.

'How much?' I asked, as he went by carrying the money in his hat.

He grinned happily. 'Enough to get me a lot of pussy in Yarmouth. About 3,000 British pounds I guess!'

During the morning the wind moderated to gale force 8, and although the seas were still spectacular, the men were able to work on the crane. It took several days of laborious work before the job was completed, and it nicely coincided with the gale blowing itself out. A ridge of high pressure drifted down from the north, giving a spell of frosty nights and sunny days. Most important of all, the sea was calm.

Within minutes of the crane's being pronounced safe, the barge carrying the engine-room package was towed in and the slings hooked on. At nearly 300 tons, it was a good test for the crane, but there were no hitches as it was hoisted into the air and swung into place against the drilling works. When the accommodation package was made ready for lifting, even the hardened platform-builders were excited at the prospect of another successful job nearing completion, and they eagerly attached it to the crane. Compared with the drilling works, at 200 tons, and the engine-room, at 300, the accommodation package was king size: it weighed close to the crane's maximum lift of 500 tons. There was an anxious moment or two when the package started to swivel as it rose off the barge, but long ropes had been attached, and men on the platform pulled it straight while it sank into place. Teams of welders, riggers and electricians commenced the task of joining up the miles of pipes and cables needed to make the rig operational. The final piece of the jigsaw to be fitted was the drilling derrick, and after the dead weight of the packages the great crane swung the metal pylon up to the drilling works almost contemptuously and held it while the riggers bolted it into place.

For several nights the blue flashes from the arc-welders lit up the platform like a firework display, then, on a warm, memorable afternoon in mid-November 1967, I was told to pack my bag and move onto the brand new drilling rig, Hewett "A". Stepping across to the metal catwalk of the rig from the deck of the *Hercules,* I paused for a moment and stared with ineffable excitement at the towering metal island, its tall spire ablaze in the winter sun. Little did I know that, almost a year to the day, I would be leaving it in less happy circumstances.

Stewards from the catering contractors were already on duty, and I was taken to the radio-room on the top floor of the accommodation block beneath the helicopter landing-deck. No expense had been spared on equipment, and when I opened the door, I was confronted with an impressive array of the latest in high-powered radio transmitters. Whoever had planned the installation was aware that a radio-operator spends his working day sitting on his backside and yet has to move quickly from one radio set to another. In the centre of the room there was a well-padded armchair which could be moved about on castors, and tied to it was a large label with the message: 'Confucius say: soft chair - sign of contented arse.'

Adjacent to the radio-room, my cabin was a spartan cubicle with a bunk on each side and two metal storage lockers. Over one of the bunks was an extension speaker from the radio-room which I knew would at least guarantee that I would have the cabin to myself. The system has been changed since safety regulations were tightened, but in those early days each rig employed only two radio-operators, usually working a rota of two weeks on and two weeks off. During his two weeks on the rig, an operator was on call at any time, day or night. Once I was attuned to the routine, even when I was asleep there was a part of my subconscious which seemed to stay 'on watch' and sound an alarm if the name of the rig came through the cabin speaker.

The rig had all the bewildering newness that most people experience on their first day in a new job, and I sat in the radio-room and waited for the 'bosses' to arrive and introduce themselves. The room was rectangular, about ten feet wide by sixteen feet long, and doubled as an office. Opposite the door, on a wide bench the full width of the wall, there were high-powered radio telephones for contacting ships or making telephone calls to anywhere in the world, VHF transmitters providing a direct link to the Phillips base in Great Yarmouth, and a special aircraft frequency transmitter for speaking to helicopter pilots and the helicopter base ashore. To assist pilots to find the rig in darkness or poor visibility, there was also a radio beacon which continuously transmitted the rig's call sign. Switching on the power, I tested the main sets by calling Humber Radio, fifty miles away, and informing them of the rig's position and that it was now operational.

I had called the Phillips base and Bristow helicopters in Great Yarmouth and was busy winding up the large brass wall-clock and setting the time when the door was pushed open. A tall, grey-haired man wearing thick horn-rimmed glasses barged in, grasping a canvas holdall in one hand and leather briefcase in the other. Without speaking, he flung the bag and briefcase onto a side bench, flopped heavily into a chair, lit a cigarette and blew a cloud of smoke towards me.

'Where's the toolpusher?' he demanded. It was an American expression, new to me, and I wondered if it was some sort of wrench.

'The what?' I said, stalling for time and hoping he would explain.

'The toolpusher!' he snapped. 'Goddam it, don't you know what a toolpusher is? Who the hell are you anyway?'

'I'm Bob Orrell, the radio-operator,' I said, holding out my hand, but he ignored the gesture.

'Holy shit, a goddam Limey radio man!' he sneered, and ramming the cigarette into an ashtray, he stalked out of the room.

I stood open-mouthed with surprise at the outburst. I had been on the rig less than an hour and the first American I met was nasty and aggressive. It was unbelievable. I wanted to go after him and tell him what I thought of ill-mannered Americans but I checked the impulse, and later I was glad I did. When I got to know him, he was one of the kindest men on the rig.

Close to the radio-room was a door leading outside to a metal grating, and I went out to check the antennas and a bank of batteries which provided emergency power for the transmitters. Visibility was superb, and I could make out every detail of the structure of the Arpet platform a few miles away. To the west, the tall chimney of Great Yarmouth power station was the only visible feature peeping above the low black line of the Norfolk coast. Trawlers were scattered over the sea like a flock of gulls, passing as close to the rig as possible, as if determined to ignore the multi-legged monster which had sprouted on their traditional fishing-grounds. I was loathe to leave the lovely view, but I heard a voice in the radio-room, and when I returned I found my chair occupied by a well-built man reading a newspaper. He swivelled the chair round as I entered, looked me over for a few seconds, then with a warm smile held out his hand.

'Hi, Bob, glad to meetya! I'm Vernon Andrews, the toolpusher. Sorry I wasn't here when you came off the barge, but I've been kinda busy.'

We shook hands, and I pulled up a chair. 'Things'll be kinda quiet for a coupla days till we get a crew of rough-necks and roustabouts sent out, so it's a chance for you to take a good look round.'

I found the jargon completely baffling and had to confess that I had never worked on an oil rig before and had no idea what a toolpusher did or what roughnecks and roustabouts were.

He looked at me incredulously and scratched his head. 'Holy smoke,' he exclaimed, 'you sure are a greenhorn, but you'll soon learn I guess. The toolpusher's the guy in charge of the drillin', and the sucker who takes the rap when the shit starts to hit the fan. That's me when I'm doin' my tour, and Mitch Green who takes over when I ain't. Roughnecks are the guys who work with the driller connectin' up drill pipe, screwin' on drill bits and anythin' to do with the drillin' operation. On the monkey board, that's the platform near the top of the derrick, there's a guy called the derrickman who stacks the drillpipe as it's pulled outa the hole, or makes a stack of pipe ready for connectin' up when they're drillin'. Then there's the rig mechanic. I guess you could say he's the same as the chief engineer on a boat. He's in charge of the engines and machinery. He's got an assistant and there's an electrician who's responsible for the generators and domestic lightin'. Roustabouts are what in England you'd call general labourers, they work with the crane-operator unloadin' workboats and doin' maintenance. The petroleum engineer will be comin' out by chopper tomorrow. He's the Phillips representative who reports back on the drillin' schedule and is in overall charge of the rig.'

The small group of men who gathered in the mess room for the evening meal were mostly American, and feeling rather out of the conversation, I climbed the steps to the helideck to watch the sunset. The flat calm sea was a deep blue, but as the winter sun dipped behind the Norfolk coast it left an inky blackness where the sea and the sky blended into one. Tiny blobs of red, green and white pinpointed the trawlers, and the brilliantly lit towers of the Hewett and the Arpet pierced the night like fiery beacons.

My first full day on the rig began with a warning on the early morning shipping forecast of a severe south-easterly gale. The *Hercules* must have heard it too, for shortly afterwards the giant jib was lowered, the anchors were heaved up and the barge, pulled by its tugs, set off in the direction of Great Yarmouth. They had nothing to wait around for. Their work on the Hewett was finished and they deserved a break. Curly, no doubt, would be cursing and swearing about the useless men he had working for him, and belly-aching about the mean, money-grabbing company he worked for, but when the barge reached the next job he would be there directing operations, and the same men would be happily working for him. Theirs was the most arduous job imaginable, and they lived hard and played hard.

As a small boy, helicopters had always fascinated me, but the nearest I had ever been to one was when I once helped to lift a mountaineering casualty into a Royal Air Force helicopter near my home in the Lake District; so it was very exciting on my first morning on the rig to get a call from Bristow helicopters in Great Yarmouth with the message that a Westland Whirlwind with six passengers on board was on its way to the Hewett and would be landing in twenty minutes. Would I make sure that there were no cranes operating and that the landing-deck was clear? And would I switch on the beacon? I pressed the switch which activated the beacon, checked that it was transmitting and climbed the metal companionway outside the radio-room to the helicopter landing-deck. It was a flat platform about forty feet in diameter, and clear except for a pair of starlings that had taken the wrong turning over Norfolk and dropped exhausted onto the rig. I had read somewhere about the danger of birds being sucked into aircraft engines, and tried to shoo them away, but they refused to budge. Each time I tried to catch them, they hopped out of reach, and in the end I gave up and went back down the steps. The cranes were less of a problem.

'That pilot ain't got a thing to worry about,' said Vernon wryly. 'The crane-operator's right there in the whirlybird with him.' Then he added, 'And if he ain't, he's fired!'

About the time I spotted a black speck in the sky approaching from the south, the aircraft-frequency radio crackled into life. 'Hewett "A", this is Golf Bravo Charlie Alpha, landing in five minutes.' I was still worried about the starlings and called the pilot to ask if they were a danger. His reply was clipped and precise. 'No danger! If still there when I land, suggest you add starling pâté to the rig menu.'

As the tiny silver-and-red helicopter circled the rig, I noticed that instead of wheels it was fitted with floats like a seaplane, and the noise of the engine and whirring of the rotors was deafening as it hovered over the helideck for a moment, then landed effortlessly on the large H painted in the centre. A door opened, a ladder dropped down and six passengers emerged, ducking instinctively to avoid the rotor and gripping their hats to prevent their being blown into the sea by the slipstream. A few items of freight were lifted out, the door closed, the 'thumbs up' was given to the pilot, and as the rotors speeded up, it rose into the air and flew away. The whole operation had taken no more than three minutes.

The passengers dispersed to their accommodation, and I was on my way from the galley to the radio-room with a mug of coffee when one of the new arrivals came out of the Phillips Petroleum engineer's cabin in what the crew later christened 'the executive suite'. (On the same landing as the radio-room and my cabin were cabins for the tool-pusher, the Phillips Petroleum engineer, a geologist and visiting VIPs.) He was probably in his forties and was dressed in neatly pressed green overalls with the Phillips '66' emblem prominently displayed on his left breast pocket. In his socks he must have been over six feet tall. The clean-cut face had a malevolent expression and a hard mouth.

I had almost reached the radio-room door when he shouted 'Hey, you!' I kept on walking. He shouted again. 'Hey you, goddammit!' I turned and looked at him.

'Are you by any chance speaking to me?' I said slowly.

'Yeah,' he glowered. 'You deaf or just stoopid?'

I took a deep breath. 'My friend,' I said, 'I don't know who you are or what you are, but there is only one way I react to ignorant bastards like you and that's to ignore them.'

It was like waving a red rag at a bull. He stormed after me, his big frame shaking with rage.

'Listen, wise guy,' he snarled menacingly. 'I'm the company man representing the owners of this rig; and when I say jump, you jump!'

'Your company may own the rig, but they damn well don't own me,' I retorted angrily, 'and if you want someone to jump for you, you'd better buy yourself a puppet!'

He seemed about to launch himself at me when Vernon appeared at the door, and with a snort he brushed past him and went out.

'What in the hell was all that about?' queried Vernon.

'That was our big friend throwing his weight about,' I said, striving not to let him see that the clash had deeply shaken me. 'And I tell you now, Vernon, if that's the type of person I'm going to be stuck out here with, you can stuff your job! I'm not prepared to be treated like shit by anyone. When the next helicopter comes in, I'm on it!'

'Calm down, Bob, calm down,' urged Vernon. 'Don't let the guy bug ya. Some of the company men like to put the boot in from the start and let everyone know who's payin' the wages.

He's just a bullshitter; there's some real nice guys working for Phillips.'

A busy hour working the radios helped to take my mind off the fracas, and about mid-morning the captain of the German supply boat *Inkenturm* called to say that he had twenty passengers on board and several tons of diesel oil, fresh water and chemicals and that he would arrive at the rig at 1300 hours. I acknowledged the call and left a note in Vernon's cabin.

The sky had clouded over shortly after the helicopter departed, and a stiff easterly breeze stirred up white caps on the sea and warned that the gale was not far away. On deck the crane-operator was busy preparing the long, flexible water pipes and testing the two cranes mounted precariously on tubular columns on each side of the rig.

Shortly before one o'clock the wind increased to near gale force, and when the *Inkenturm* came into view it was crashing heavily through the seas, throwing clouds of spray high in the air. It seemed impossible to contemplate landing men in such conditions, but Vernon did not look at all perturbed and yelled at the crane-operator to hook the personnel basket on. This device was shaped like an inverted shuttlecock, with a lifebuoy about six feet in diameter at the base and rope netting at the sides tapering to a small ring, which was attached by a short rope to the hook of the crane. Baggage was thrown into the centre, and men, optimistically wearing life-jackets, stepped on the large ring and clung to the net while the basket was hoisted high over the sea and swung inboard to the rig.

'I haf four passengers very seasick. You please can take them first,' said the German captain over the radio. 'I cannot tie up, conditions are not good, but I vil kom on your vest side.'

The radio-room was on the west side of the rig, and I had a grandstand view as the captain manoeuvred the boat underneath the crane while the basket was lowered towards the heaving deck. One minute the deck would be level with the basket, then, as the boat sank into a trough, it was ten feet below it. A long rope trailed from the base of the basket to enable the boat's crew to hold it and stop it swinging about but, even so, getting the men to stand on it and grab the net was fiendishly difficult in the sea conditions. For a reason known only to those who financed the rig, portable hand-held radios had not been supplied, and the only way the crane-operator could communicate with the boat was by shouting or using hand signals. In a howling wind and thundering seas, neither was very effective.

Several attempts were made to lift the men but each time, before the crane could winch them clear, the boat rose sharply on a wave crest, hitting the basket and throwing the men to the slippery deck. Throughout the operation the captain raged at me over the radio to get the crane-operator to hurry, and I did my best to explain that he was doing all he could. Eventually four men and baggage were plucked from the deck and dumped unceremoniously on the rig like a heap of half-drowned rats. Their faces were green with the effects of the sea-sickness, but there was no sympathy.

'OK, you guys,' said Vernon, 'get your asses down to the stores and get fitted out with coveralls and hard hats. You're on the twelve till midnight tour.'

They looked up from the heap in shocked disbelief, then with loud groans vomited over the new paintwork.

By the time the rest of the men had been lifted aboard, there was no question of the boat's discharging fuel and water. It was being tossed about like a cork and was in danger of colliding with the rig.

'Tell that skipper he can pull away,' bellowed Vernon above the noise of the gale, but when I looked through the window, the German was already heading off at full speed.

With a crew of forty on the rig, the shifts - or tours, as the Americans called them - settled into a routine. For a time, the British contingent, including me, had difficulty in understanding the strange accents of the Southern States and the drilling man's jargon. In the drilling business time was expensive, and the vocabulary of the men was as economical with words as the companies were with their money. There was no room for gracious expressions like 'please' and 'thank you', and our robust Anglo-Saxon four-letter words paled into insignificance against the obscene character-assassinations that were part of the American oilman's everyday language.

Ordinary words like faucet for tap, handsocks for gloves, slicker suit for waterproofs, john for toilet, vest for waistcoat, pants for trousers, life-vest for life-jacket, and lines for pipes that carried diesel, water or cement, caused a lot of confusion. Nothing was ever dismantled: it was 'torn down'; and 'hit the switch' was an order not to thump it with a hammer but merely to switch it on. The belligerent way in which some Americans spoke to the men working under them sparked off many heated disputes, with fistfights and broken jaws on both sides, before the British were able to understand that more often than not there was nothing personal in it. It was the law of the oilman's jungle, where, as in any other jungle, the animal that was most aggressive and made the loudest noise was respected by the leaders of the pack and rose rapidly in the pecking-order.

The lure of the rigs attracted men from all walks of life and of many different nationalities. The Hewett crew list read like a gathering of the United Nations, with men from Britain, the United States, Canada, Greece, Turkey, Morocco, South Africa, Australia, Spain, France and Pakistan. Some were fugitives from the law, a few had heart-rending personal problems, many were students working their way round the world who worked for one or two weeks, then were gone. There were well-educated businessmen who had gone bust, drug-addicts, men fighting a drink problem and those who, for one reason or another, had opted out of society. As I got to know them and they grew to trust me, the radio-room became more than a communications centre. It was a confessional, a lonely hearts' club and a citizens' advice bureau.

The gale gusted to over seventy miles an hour during the next couple of days, and the violent seas gave the rig a severe pounding, flinging spray against the radio-room window eighty feet above sea-level. At first it was a disturbing experience to feel hundreds of tons of steel, perched on stilts, swaying slightly like a tree, but Vernon explained that it was a safety factor and that all fixed platforms were designed to 'give' a little under wind and sea pressure. The *Inkenturm* rode out the storm, alternating between steaming slowly into the wind and drifting back, and when at last the wind moderated and the sea subsided, the skipper brought his boat alongside to discharge the diesel oil and water.

Following the Sea Gem tragedy, the British Government insisted that there must be a boat standing by every offshore rig twenty-four hours a day, and it provided welcome work for the redundant trawlers of the beleaguered British fishing industry. As the *Inkenturm* departed for Yarmouth, the stern trawler *Ross Delight* arrived at the rig and anchored about half a mile away. When I talked to the skipper on the radio, he was glad to be in work but unhappy at the prospect of rolling at anchor by the rig for several weeks after the excitement of deep-sea trawling.

'It's an old man's job this,' he complained, 'but I suppose anything's better than walking the bleeding streets of Hull and getting pissed out of your mind every night with boredom.'

After lashing the rig with a storm, the weather went to the other extreme, with no wind, a flat calm sea and fog so dense that visibility was no more than a few feet. Walking anywhere on deck was very hazardous, with the risk of misjudging a step and falling down a companionway. To add to the worry, the electric foghorn on the rig died after emitting a few asthmatic grunts, and no amount of electrical or mechanical resuscitation, or the shouting and bawling of the Phillips company man, could revive it.

'That cocksucker makes more goddam noise than any fog-horn,' grumbled the Kansas City rig mechanic as he bolted the inspection plates into place. 'All we need is the sonofabitch to stand on deck and holler and no boat would come near us.'

Laughing at the idea, everyone agreed that it would be very effective, but Vernon was taking no chances and said I should call the skipper of the 'fishboat' and tell him to keep a good look-out for approaching ships on his radar.

It seemed as if the fog would never lift, and when the time arrived for me to go ashore for my two weeks off, all helicopter flights were grounded and the *Inkenturm* was despatched from Yarmouth with the crew-change. When it nosed through the fog alongside the rig, the sea was so calm that the crew were able to step from the boat landing onto the metal stairway used by the rig builders. Rimrock Tidelands had appointed a relief radio-operator, Alan James, an ex-Royal Navy signaller from Newcas-tle, but there was hardly time to shake hands and show him the radio installation before the onshore-going party swarmed on board the boat and the captain was sounding his siren, signalling that he was about to leave.

It was a tedious, uneventful passage at slow speed, with nothing to see but grey fog, but when we were about ten miles from Yarmouth the youthful captain came down from the wheelhouse, beaming with the news that one of his company's supply boats had reported that the fog was lifting further south and visibility was perfect. Within half an hour we had steamed out of the fog, and ahead was the entrance to Yarmouth harbour and the town bathed in sunlight, with the spectacular Nelson's column, built long before the London version, silhouetted against an azure sky. The *Inkenturm* slid between the pierheads, rounded a bend into the main harbour and tied up at the Phillips Petroleum Company base.

To my annoyance, we were told we could not leave the boat until we had cleared customs.

'Why on earth do we have to go through this rigmarole?' I asked the brusque and unsmiling customs officer as he searched through my bag. 'We're hardly returning from a Continental holiday!' He was not amused, and showed it by demanding to see my passport. 'I haven't got a passport,' I said irritably. 'What the hell do I need a passport for? I work on the Hewett "A" rig, which is only fifteen miles off the Norfolk coast; that's hardly going overseas.'

'If you go outside territorial waters, you are deemed to have gone foreign and you are required to be in possession of a passport,' he said, and pulled out a notebook from his pocket. 'Can I have your name and the company you work for?'

None of the Hewett crew on the *Inkenturm* had passports, and over an hour was wasted before the officious duo had satisfied themselves that we were not a gang of mercenaries intent on taking over the country, and allowed us ashore. It was typical of inflexible British bureaucracy that they were unable to appreciate that, as the offshore drilling and production activity was expanding on the British Continental Shelf, men whose movements were well documented by the Company they worked for were simply going out to, and coming back from, the same rig. The Government argued that there was no case for revising customs procedures, yet, when it suited them, they always seemed able to alter the rules in their favour. Some time later, when it was requested that men working on rigs outside the three-mile territorial zone should be allowed duty-free tobacco, cigarettes and cigars, the Customs and Excise refused permission on the grounds that they were not working outside the UK.

The harbour was alive with activity. Supply boats churned the muddy waters of the River Yare, and pick-up trucks bearing the emblems of world-famous names in the oil industry darted in and out of the brightly-painted new compounds. Basked in the bright light of the afternoon sun, it could have been any port along the coast of Texas or Louisiana, but the empty cigarette packets, greasy chip-wrappers and discarded copies of the *Daily Mirror* littering the streets shattered the illusion.

A busy east-coast port for centuries, Yarmouth was badly affected by the decline of the fishing industry, and though the harbour was still very much the heart of its town, it was the sandy beaches along the coast, not the sea, that provided its life's blood. The boom in tourism after the 1939-45 war gave Yarmouth a new lease of life as a seaside resort, and the new visitors demanded accommodation and entertainment. Leisure complexes sprouted up along the sea-front, and waste ground was cleared to provide vast caravan parks. Hotels flourished and the rows of large Victorian houses displayed bed-and-breakfast or guest-house signs.

Shipping continued to use the harbour but went largely unnoticed - the big money was in candy floss, bingo and letting-accommodation. As far as the business community was concerned, there were no fortunes to be made from the harbour; the tourist ruled.

The first probes for gas under the North Sea were mentioned only briefly in the British press, and the unusual craft arriving at east-coast ports received only scant attention. It was not until the civil-engineering company of George Wimpey formed a partnership with Brown & Root of Texas and converted an old fish-curing plant into a North Sea exploration shore-base that other companies were attracted to the port, and Yarmouth was launched into the most exciting enterprise of the century. It had everything the explorers needed; not least, plenty of unused buildings, space and a harbour that was accessible at any state of tide. From then on, the pace of the development of Yarmouth was almost meteoric. American oil companies, drilling companies, shipping and construction companies, together with dozens of service companies from the States and Europe, flooded into the town, eager for bases and accommodation for staff - and with unlimited money to pay for them. Thousands of feet of rotting quays were hurriedly rebuilt, and silos and bulk storage tanks sprang up almost overnight. Land- and property-owners abandoned the seasonal tourist trade, and caravans were cleared off sites to reap a more lucrative income from renting the space for pipe yards, machinery stores and office accommodation. House prices rocketed and rented property soared far out of reach of local people. Americans complained that information supplied to them by the British Government was misleading, in that it had stated that they could 'live like a king' in Britain for £150 a month. When they reached Yarmouth, they discovered that the monthly rent demanded for even a small house was almost that figure.

Many local people argued bitterly that the new prosperity was artificial, transitory and damaging to the long-term future of the town, but few would listen. Those who owned property in the town were sitting on a fortune, and for the 300 applicants a day who besieged the offices of the exploration companies, the prize was a well-paid job and an opportunity to get a toe in the door of an industry that showed every indication of being around for a long time. Of the eleven rigs operating in the North Sea in 1967, nine were supplied and serviced from Yarmouth; and in two years the figure for ships arriving in Yarmouth in the month of March rose from 165 to nearly 600. The spin-off benefits were enormous, and provision merchants, restaurants, hotels, car-dealers, laundries, taxi companies, travel agents, welding companies, engineers and anyone with a service to offer could soon have a very contented bank manager. But first a few hard lessons had to be learned.

To many owners of businesses in Yarmouth, the word 'service' meant taking their car to a garage to be greased or attending a church meeting on Sunday. It certainly never meant working after office hours and being on call twenty-four hours a day, seven days a week. For them there was to be a great awakening.

While I was at home, the Rimrock office secretary phoned to say that I was to report to Bristow's heliport, and not to the harbour, on my return. At the end of my two weeks' break I drove to Yarmouth eagerly anticipating my first flight in a helicopter.

The cluster of buildings in a corner of Caister airfield that formed the helicopter base had been described by one enthusiastic newspaper reporter as 'an international airport in miniature', but the only thing it had in common with Heathrow was the foul-tasting hot drinks dispensed in plastic cups from slot-machines in the passenger terminal. Checking in at the reception desk, I sank into a chair and sipped a cup of brown liquid the machine claimed was coffee.

Bristows had won most of the contracts for transporting crews to the rigs, and almost from first light the dawn peace was shattered by the whirring roar of helicopters taking off and landing like wasps at a nest. The small terminal building overflowed with men bearing the stickers of a dozen different companies on their luggage, and British workers, bleary-eyed and weary after an all-night journey from the north of England or as far away as Scotland, sprawled fast asleep across their luggage on the floor. Swarthy catering-company stewards giggled over a girlie magazine and, though the comments were in Arabic, the gestures and sound-effects needed no interpretation. A trio of Frenchmen, arguing noisily about the political situation back home, gesticulated with Gauloise cigarettes and blew clouds of acrid smoke into the air to mingle with the fragrant aroma of cigars gripped in the teeth of stetson-hatted Americans.

As the name of a rig was called by the receptionist, cigarettes and cigars were hastily stubbed out, the last dregs of beer drained from cans, half finished cups of coffee abandoned and prostrate figures roughly shaken into life as the rig crew shuffled into an ante-room to pull on full-length survival-suits and Mae West life-jackets. This uncomfortable combination was insisted upon by the Air Ministry in the pious belief that, if a helicopter ditched in the sea, those on board could splash about happily in the water until help arrived. As ever with Whitehall wonderboys, they overlooked the fact that, to be effective, a survival-suit had to be a good fit, and since helicopter operators could hardly be expected to cater for every variation of the human form, it often happened that a beefy American walked out to an aircraft looking as if he were modelling the latest in stretch rubber and followed by what appeared to be a caricature of a Michelin man.

The Hewett crew were called and, dressed in our blue survival-suits, we clambered onto the floats, stepped inside the ten-seater Westland Whirlwind and wedged ourselves into the narrow seats. It was icily cold after the warmth of the passenger terminal, and I shivered inside the suit that was to protect me if I was thrown into the sea. There was no mini-skirted air hostess offering in-flight refreshment or entertainment. The pilot came aboard and gave curt instructions that if we ditched into the sea we were not to inflate life-jackets or the life-raft until we were out of the helicopter; smoking was forbidden and safety-belts were to be kept fastened. Then, closing the door, he went outside to climb into the cockpit. For a few minutes nothing happened, then, with an ear-splitting whine, the engine burst into life, and as the large rotor started to revolve, the whole machine vibrated and rattled as if it was about to disintegrate onto the tarmac in a heap of nuts and bolts. The eardrum onslaught rose to an almost unbearable pitch, dropped for a few seconds, then rose again as the helicopter lifted off the ground, spun round in a half circle and flew above the runway for a short distance before climbing into the sky. Through a window I watched the passenger terminal, aircraft hangars and parked helicopters shrink to the size of small toys on a painted board as the ground fell rapidly away. Banking steeply, the helicopter turned high above the lines of parked caravans and a long sweep of sandy beach and headed out over the grey sea.

Conversation was impossible without shouting above the noise of the engine, and the passengers flipped through magazines or gazed through the side windows at the constantly changing picture below. A few miles offshore a long convoy of trawlers was heading home to the port of Lowestoft, a few miles south of Yarmouth, at the end of a fishing trip. Having worked on trawlers, I could imagine wives and girlfriends awaiting the arrival of their men with apprehension, and pub-owners polishing their glasses in joyful anticipation. There was a stir of excitement when we flew over a jackup oil rig being towed north by two tugs, and the pilot must have wondered what was happening to his controls when, unseen by him, the passengers dodged from one side of the helicopter to the other, trying to make out its name.

Looking down at the grey sea, there was no way of judging speed and distance, and I was not aware we had arrived at the Hewett until the engine revolutions suddenly dropped. I looked out to see men working on deck, and felt a slight bump as the helicopter landed. The rotor was kept running while we darted across the deck and down the steps into the accommodation and handed over the survival-suits to the crew going ashore. Again, there was no chance to exchange more than a few words with

Alan, the second radio-operator, and by the time I had dumped my bag on my bunk, collected a cup of coffee from the galley and slid into my seat in the radio-room, the helicopter was well on its way back to Yarmouth.

Vernon Andrews was one of the crew who had gone ashore on the helicopter, and when I discovered who his relief was my heart sank. It was Mitch Green, the tall, grey-haired man who had been so unpleasant on my first day on the rig.

'Hello,' I said cheerfully, when he came into the radio-room and rummaged through a desk drawer. 'Enjoy your week off?'

Completely ignoring me, he sat at the desk filling in the drilling records, then replaced them in the drawer. He snapped it shut and without saying a word went out.

'Well, stuff you,' I muttered after him, and turned to answer a ship calling on the radio.

Much had happened while I had been away. The drilling of the first well had been started, or 'spudded in' to use the drillers' expression, and the rig shook with the rumble of machinery and the clanging of steel pipes that went on unrelentingly, day and night. There had been an angry exchange of words between the Americans and the catering staff when, on the third Thursday in November, the chef in his innocence cooked the usual meat dishes for the main meal. Unknown to him, it was American Thanksgiving Day and should have been celebrated in the traditional way with roast turkey and all the trimmings of a festive occasion. The Americans took it almost as a personal insult when they sat down to roast beef and Yorkshire pudding and were so incensed that, to appease them, the real thing was specially flown over from the States, complete with a large quantity of soft ice-cream.

I saw very little of the taciturn toolpusher, and if he spoke to me at all, it was in incomprehensible grunts which, because I could not understand a word he was saying, made him even more antagonistic towards me. It seemed we were on a collision course whichever way we turned, and though I tried hard to be friends, he made it clear that British radio-operators were not his flavour of the month. He did his damnedest to be awkward and unco-operative, but about mid-week he lifted my flagging morale with the news that the large and loud-mouthed Phillips Petroleum engineer had been transferred to a project in the Middle East. His place was taken by Bill Macmillan, a mild-mannered, quiet-spoken Texan, but one who could be as hard as nails if the need arose. I took an instant liking to him, and he taught me a great deal about drilling and the oil industry.

Provided all was going well, the relationship between the Phillips man and the Rimrock toolpushers was one of mutual tolerance in order to get the job finished, but it was one of the Phillips engineer's responsibilities to see that his company got its pound of flesh, and if things started to go wrong, the veneer of friendship melted like an ice-cream in the Sahara. In one particularly nasty shouting-match between Bill Macmillan and Mitch, I feared they were about to come to blows, but it ended with Mitch storming out of the radio-room and Bill angrily kicking the door shut behind him.

'He's as stubborn as a goddam fucking mule!' raged Bill, more to himself than to me, and tearing a few sheets off a notepad he stuck them on a clipboard with a sheet of carbon paper and wrote a furious report to his office. Having got whatever was bugging him out of his system, he leaned back in his chair and rolled a cigarette.

'You need to be on the ball in this game, Bob,' he confided, striking a match on the heel of his boot and sucking the flame into the end of the cigarette. He inhaled deeply and let the smoke trickle out of his nostrils. 'Make a mistake in the oil business and you can go from hero to zero quicker than you can spit, and it don't matter a damn what good work you've done in the past. One thing you learn pretty quick is that, if a company ever pats you on the back, all they're doin' is lookin' for a soft spot to put the knife in.' Apart from the occasional stroll round the deck I had seen very little of the rig and, taking advantage of a lull in the arrival of ships and helicopters, I asked Bill MacMillan if he would sit in the radio-room and listen out while I took a wander round.

About 300 feet long and 150 feet wide, the platform lay roughly north to south, with the drilling derrick on the north end.

The large accommodation block at the south end was topped by the helicopter landing-pad, or helideck as it was generally referred to, and in the centre was the engine-room block, whose top provided deck space for the storage of drilling pipe, casing and pallets of chemicals. On the east and west sides of the engine-room, vertical columns supported the cranes and also served as tanks for the diesel fuel. Between the north end of the engine-room and the derrick was a wide gap crossed by a slippery gangway, like a drawbridge over a moat. On either side of it there was a long drop into the sea. It was a place notorious for accidents, and with the usual American flare for descriptive realism it was known throughout the drilling world as 'the widow-maker'. It provided access to the drill floor for the drilling crew, and casing and drill pipe was winched across it into the derrick. The 130 foot-high derrick rested on massive girders which also supported the drill floor, the nerve-centre of the drilling operation. Below the platform under the derrick was a series of hazardous catwalks which supported the blowout-preventer, a clever device designed to close round the drill pipe in an emergency and stop gas escaping. Under the platform on the south end of the rig a metal stairway zigzagged down to boat landings on the east and west sides at sea-level.

One of the most fascinating, dirty, wet and dangerous places on the rig was the drill floor. The jealously-guarded preserve of the driller and his team, it was almost sacred ground, and when I strayed onto it, curious to see how drilling was carried out, a large hand gripped my shoulder and above the squeal of machinery a voice bellowed, 'Say, who the hell are you? Get your ass off my drill floor pronto!' There was no arguing with a Canadian built like the side of a skyscraper, and I beat a retreat; but when I eventually got to know him, he was a great help, and even at his busiest he was always willing to explain what he was doing.

Men who had spent their lives in the oil industry lived it, breathed it and talked about nothing else, but drillers were a race apart and were as hooked on their work as an addict is on drugs. Their one consuming passion in life was drilling holes in the ground, whether it was dry land or the sea-bed. What little conversation they had was about drill-pipes, casing, rotary bits, joints, tongs, kellies and mud. They might curse and rave about the company they worked for or the conditions they had to work in, but after only a few hours ashore they longed to be back on the rig. For a driller, happiness was the whirling rotary table controlling the drill bit that cut into rock thousands of feet below ground. 'Drillin' hole' was a disease with no known cure.

The method of drilling for oil or gas which has evolved from the early American experiments is now more or less the same anywhere in the world, and it was Bill Macmillan, during our pow-wows in the radio-room, who helped me to get a clearer understanding of what was happening underneath the rig.

'Folks maybe see rigs on television but they ain't got a cotton-pickin' idea where oil or gas comes from,' Bill said, biting the end off a big cigar and spitting it into a wastepaper bin. Lighting up, he blew a smoke ring towards the ceiling. 'Yessir, they think it's trapped in a big underground cave and if you drill into it, out gushes oil or gas. But it ain't like that. The rock way down there where there's gas is porous, like maybe a hard sponge. It's covered with thousands of feet of solid rock. The gas can't get through, and it stays there until one day a rig sits right over it and drills a hole maybe nine or ten thousand feet down to the sponge. Then, boy, you'd better watch out, up she blows! In the old days they used to let it gush out like you see on the movies when they strike oil, but things have kinda improved since then. We plug it before it gets outa hand.'

He paused to pick up my notepad and, taking a pencil from a desk drawer, drew a rough sketch of a rig and a series of lines underneath it representing various depths.

'All the different names are kinda bafflin' till you know your way around, but I c'n give you a basic idea. You've got to imagine a column of steel pipe with a drill bit on the end of it being spun round - rotary drillin', we call it. The weight of the pipe on the bit makes it chew away at the rock until it's bored a hole nearly the depth of the column. Then a second column of pipe is screwed on, and it spins away again until another column has to be screwed on; and the whole process keeps on till you've drilled down thousands of feet. That's roughly how it's done, but there's a refinement or two to stop the whole string of pipe fouling up in the hole.'

The cigar had gone out, so he lit it again, made sure the end was glowing and drew two vertical lines on the note-pad representing a drilled hole.

'If that was just a hole drilled in the ground, the sides could collapse and jam the pipe, so what we do is drill the first couple of hundred feet with, say, a twenty-inch bit. Then we sink steel casing down the hole and cement it in. On an offshore rig like this, the casing would be raised well above sea-level to keep the water out of the hole. OK so far?'

I studied the sketch and nodded.

'Then a hole a mite narrower than twenty inches is drilled and more casing run in and cemented, and all the way down to the bottom of the hole the diameter of the bit is reduced, until you end up with a steel tube starting at twenty inches diameter at the top and tapering to maybe seven inches at 10,000 feet down. Course, all that drillin's gonna heat the drill bit up, and that's where the other refinement comes in. We pump mud down the centre of the drill pipe and out through holes in the bit. The mud cools the bit down, then works its way back up between the drill pipe and the casing, helping to cut down on friction and bringing up rock samples to the rig, so the geologist guy can tell what's goin' on down there. Course, the mud we use ain't just any old mud you'd find on a beach. In the old days when they were drillin' on land, they use to dig a pit and pour water in and mix it with soil - that's how it got the name mud - but mud ain't just there to cool the bit. No, sir, it's the weight of the mud pumped into the drill pipe under pressure that keeps the gas or oil from blowin' outa the hole. The heavier the mud the better, so we use a mixture of different chemicals we call "weight material". It's mighty simple and mighty effective!'

He balanced the cigar on the edge of the desk and held up the finished sketch at arm's length to admire it.

'Jeez!' he exclaimed with a big grin. 'Maybe I could get me a job as a professor at the state university. I'd sure enjoy teaching those pretty gal students a thing or two.' Glancing at his wristwatch, he yawned deeply. 'Well, I guess I'm gonna take a nap. We'll talk some more another night.'

During my occasional visits to the drill floor I learnt something of the practical side of drilling, and basically it was quite simple. Suspended in the derrick was an enormous block-and-tackle or pulley arrangement. One pulley, the crown block, was fixed at the top of the derrick, and the other, the travelling block, could move up and down, controlled by an electric winch on the drill floor. At the lower end of the travelling block was a swivel, and attached to that a square sectional length of pipe about forty feet long called a kelly. To the end of the kelly was screwed a length of drill pipe, and to the end of the pipe a drill bit was attached. In the centre of the drill floor, where the crew worked, there was a circular slab of metal with a hole in it, called the rotary table, which was driven by an electric motor and could revolve at various speeds. Suspended by the kelly from the travelling block, the drill bit and pipe were lowered down through the hole in the rotary table until the square kelly could be wedged. As a clutch was engaged and the rotary table revolved, it turned the kelly, which rotated the pipe and drill bit at about 150 revs per minute, and so it bored its way into the sea-bed. As more and more sections of pipe were screwed on, the hole was bored deeper and deeper until it reached its target depth of perhaps nine or ten thousand feet. The weight of the drill pipe bearing down on the drill bit helped it to cut into the rock. During the drilling, mud was pumped, under pressure, through the top of the kelly and down the hollow drill pipe to the bit.

Many of the roughnecks working with the drillers were British and had either gained experience on other rigs or shown that they got on well with the Americans and could cope with hard work and long hours.

For the crane-operators in charge of the roustabout crews, life was by no means smooth. The stories of 'big money' to be made on the rigs attracted men from all over the UK, but the tough working-conditions very soon sorted out the hard men from those who only thought they were. For a time the supply boat ran a regular shuttle service ferrying boatloads of noisy, beer-swilling, macho men out to the rig and returning with a subdued and disgruntled huddle of humanity, who had learnt the hard way that the 'big money' on the rigs was earned, not made. The British worker, accustomed to morning and afternoon tea-breaks, was rocked to his boots to discover that during his twelve-hour tour on the rig he was fortunate if he got half an hour for a meal-break. It often caused bitter resentment and was a breeding-ground for the sea-lawyers and rabble-rousers who tried every trick in the book to stir up trouble and bring the men out on strike. In an industry where men could turn their hands to almost any task and where there were no arguments about who should screw two pipes together, the same rabble-rousers fought to introduce demarcation of jobs, the very cancer that at the time was bringing British industry on shore to its knees. To waste time and sneak away for a smoke, some men deliberately threw new cans of paint and paintbrushes into the sea when they were assigned to painting.

Both the Rimrock toolpushers tolerated the troublemakers and sometimes gave in to their demands, but when they looked for something new to complain about and turned their attention to the quality of the food, the patience of the Americans began to wear thin. How they found fault with the food was difficult to understand. They gulped down chunks of prime steak like starving piranha fish, and the thick ham, sausage, eggs and mushrooms heaped on their plates at breakfast time would have satisfied a family of four.

The crisis reached a head one morning just before the midday change-over, when a particularly voluble individual called a roustabouts' meeting on deck to demand more pay. Vernon was angry and perplexed by this latest challenge to his authority but let them have their meeting.

'I sure don't know what's got into these guys,' he muttered as we listened to the man haranguing his bewildered audience. 'I ain't never given a fella a bad deal in mah 'ole life. Hell, these sonofabitches expect to get a raise when they've only done a coupla tours.'

The meeting ended with a unanimous show of hands, and the leader swaggered into the radio-room and thrust a sheet of paper at Vernon.

'It's our Three Point Plan, and if we don't get an agreement before twelve, we're on strike.'

Vernon glanced briefly at the paper, rolled a cigarette, lit it, then said, 'Sure, I'll come to an agreement. Get the men on deck. I'll have a word with 'em.'

The man could hardly conceal his triumph and, slamming the door behind him, went back to his pals. When he had gone, Vernon read through the paper again.

'More pay, less hours, better food,' he murmured. Then slowly and deliberately he dropped it into the waste bin. 'Bob,' he said quietly, 'tell that goddam boat to come alongside.'

The *Inkenturm* had arrived during the night with cargo and was anchored off the rig waiting for instructions to return to Yarmouth. Providentially the sea was flat calm, and when the captain brought his boat in, he was able to tie up against the landing. Pulling on his hard hat, Vernon walked out to where the leader and his men were waiting expectantly.

'OK, you guys,' he said loudly. 'I've come to an agreement. I've agreed with mahself that you ain't nothin' but a bunch of trouble. As of now, you ain't on the pay-roll no more, so get your asses aboard that boat. You can have your strike, but have it in England.'

The men gasped with surprise, and the leader rushed at Vernon with fists flailing, but the rig welder, an ex-wrestler from Gateshead who had been busy cutting pipes on deck, stepped forward and grabbed his arm.

'If you want to take a swing at somebody, pal,' he said softly, 'try me!'

Half an hour later the dejected strikers trooped aboard the *Inkenturm* with their baggage, and it steamed away towards Yarmouth.

'That's got rid of that heap of chicken-shit,' sighed Vernon, sinking into a chair and lighting a cigarette. But I said nothing. I felt ashamed to be British.

There was no official union representation on the rig and had the men's grievances been justified I would have supported them wholeheartedly, but Vernon was right. They were trouble-makers and the rig was a happier place without them.

Halfway through December, the Westland Whirlwind helicopter on which I was returning to the rig suddenly did an about-
turn and flew back to the airfield. We had no means of commu-
nicating with the pilot and assumed he had been requested to
return to pick up something vital for the rig; it certainly would
not have been a crew member - there was an unwritten rule that,
if a man missed a crew-change flight, and it did not matter for
what reason, he might as well head for the nearest employment
exchange, for he was out of a job. We landed at Caister with an
unusually heavy bump, and when we were asked to return to the
passenger terminal, we discovered that a mechanical fault had
developed in the rotor mechanism and it could have seized while
we were in the air. A few days later, all Whirlwinds were
grounded when a Whirlwind of the Queen's Flight plunged into
the Berkshire Downs, killing all four people on board. A defect
was later found in the main gearbox.

The grounding of the Whirlwinds caused great inconvenience
to Bristows, but when they had to resort to using the larger twin-
engined Wessex helicopter, it gave the rig crews a taste of spa-
ciousness and comfort which was a welcome change from the
noisy, cramped Whirlwinds. A shortage of Wessex helicopters
meant that, instead of making individual flights to rigs, they had
to make round trips, calling at several rigs, often involving long
and tedious journeys.

A few weeks after the drama in the Whirlwind, on the day I
was due to go ashore, I sat miserably in the radio-room listening
to a gale howling round the rig, convinced that any minute the
helicopter base would call to say that all flights were cancelled.
There is nothing in a rigman's life guaranteed to plunge morale
to zero quicker than a crew-change helicopter not arriving on
time or worse, being cancelled, but with the wind gusting to 50
mph, making it difficult even to stand upright on the helideck, I
realized that landing a helicopter would be tricky. The dreaded

call came, but to my surprise it was to let me know that the Wessex had been delayed calling at another rig and would arrive at the Hewett in half an hour. Right on time, the helicopter landed; there was a quick scramble of bodies out, then in, and we were airborne, heading for the jackup rig Constellation to drop some equipment before making for Yarmouth.

Through the window the frenzied sea far below looked daunting, and a large cargo liner, rolling heavily, was lost from view each time a great wave broke over it. Occasionally a particularly nasty squall buffeted the helicopter and caused it almost to spin round or drop viciously for a few feet and shake like a wet dog throwing water off its back.

Reading a magazine was out of the question. The motion was too violent for me to do anything more than grip the sides of my seat and steel myself for the next hole in the sky. As the helicopter dropped into it, my body and stomach felt as if they were being flung in opposite directions.

From 500 feet up, the gale-lashed frame of the Constellation looked as if it was about to be devoured by the seas at any moment, but as the Wessex descended, I could see that it was jacked high out of the water on its four latticed steel legs, and drilling was in full swing. The landing-pad stuck out from one end of the rig uncomfortably close to the giant legs towering above the deck, and to make matters worse, the fierce wind was blowing from an awkward angle. With a practised skill that was all too often overlooked in the annals of North Sea operations, the pilot manoeuvred the helicopter backwards until he was over the landing-pad, then dropped it neatly to the deck. There was little room for error. One wrong move with the controls or a moment's indecision could have spelt disaster. A few men jumped out into the teeth of the gale, and some hurriedly climbed in; but when the pilot revved the engine for take off, he seemed to hesitate, then dropped the revs again. He did this several times, then to my surprise stopped the engines, a very unusual occurrence on a rig, especially in bad weather. The door opened and a very apologetic pilot stuck his head in to tell us that an oil warning-light had flashed on and he could not take off until a Bristow's mechanic could come out to check it. There

were loud groans of dismay and shouts of 'goddam junk chop-
pers' and 'chicken-livered Limey pilots', but most of the passengers took the news stoically and gratefully accepted the rig
chef's offer of coffee and sandwiches in the galley. When the
pilot and the co-pilot returned from radioing Bristow's base, the
grim expressions on their faces disclosed that worse was to
come. 'I'm sorry, chaps,' said the pilot, 'but a warning of storm-
force winds has been issued and all aircraft have been grounded.
It means, I'm afraid, that we shall have to stay here until a mechanic can get to us.'

Streams of obscenities flowed from the American who had
accused the pilots of being 'chicken-livered', yet, when they
ignored the jibe and showed concern about his urgent need to get
ashore, it transpired that his sole interest was in a woman he had
met in Yarmouth.

'Ah sayed that the very next time ah was in town ah'd show
her that there were bigger things in a Texan's pants than his
bankroll.'

The pilot's face reddened with anger but he kept cool and
glared at the American with unblinking eyes.

'If you can't appreciate the situation, sir,' he said with delib-
erate English politeness, 'it would appear that you have more in
your trousers than you have in the space between your ears!'

Other Americans among the passengers, who were obviously
embarrassed by the Texan's comments, took the heat out of the
situation by telling him to shut his big mouth.

The rig catering manager was a very adaptable person, and
feeding and accommodating an extra ten men at short notice was
all in a day's work. He quickly allocated bunks and laid on a
marvellous midday meal.

In the early afternoon, the wind increased to such ferocity
that white spume from the wave tops was streaming over the rig
as if someone had poured thousands of tons of soap-powder into
the sea. The pilots, fearing that their helicopter might be blown
away, crawled on their hands and knees along the helideck with
three of the Constellation's roustabouts to secure it to the deck
with steel cables. On board the rig's standby boat, an ex-
Lowestoft trawler, eating or sleeping would have been impossi-

ble, and the crew must have watched anxiously as the gigantic seas, higher than the mast, rolled down, threatening to annihilate the tiny boat like a floating leaf under a waterfall. At the last minute the trawler was lifted high into the air, exposing almost three-quarters of its weed-encrusted hull, before gravity took over and it fell down the back slope of the wave, hitting the bottom of the trough with a juddering, sickening crash that hurled tons of water over the deck and wheelhouse, blotting it from view until it reared up like a surfacing whale, water pouring from its sides.

By nightfall, the storm had reached force eleven on the Beaufort wind scale, or in land-speed terms the region of 70 mph, and the seas were terrifying. I was relaxing in the mess-room reading a sailing magazine when something hit the rig with a tremendous thud, and it shook as if a ship had collided with it. There was a mass exodus from the room to look outside, but apart from the ghostlike spume sweeping over the deck under the glare of the deck lights, there was nothing to be seen. It was too wild to hang about outside. We had just returned to the mess-room when again the rig was shaken violently by a powerful blow. A few minutes later the rig toolpusher came into the room, shaking the water off his hands and dripping a large pool onto the polished floor from his oilskins.

'OK, you guys, we got a slight problem here. If ah hadn't seen it with mah own eyes, ah would have said "bullshit". We've got the barge jacked up fifty feet above sea-level, but some of those waves are so goddam big they're kicking it up the ass. We can't jack up any more in this wind, so I guess we'll have to sit it out.'

Almost as soon as he had finished speaking, another wave hit the rig and seemed to lift the legs off the sea-bed before letting it crash down again. As the evening progressed, we were being pounded by monster waves with unnerving regularity, and as a precaution everyone was told to stay awake and have a life-jacket handy. The big worry of the toolpusher was that the scouring action of the turbulent water on the sea-bed might affect the stability of the rig's legs; in extreme circumstances it has been known to cause a rig to fall over, an accident which in storm-force winds and mountainous seas did not bear thinking about. If the toolpusher thought about it, he certainly did not show it, and he never ceased to be cheerful. Throughout the night he and the rig mechanic and a few helpers regularly risked their necks on deck in the teeth of the storm to check that all was well.

When the tide turned and the tidal stream ran with the direction of the wind and not in opposition to it, the wave height decreased and the pounding gradually eased and then ceased altogether. Men went to their bunks, life-jackets were deflated and stowed away, and with an exclamation of relief the toolpusher sank into a chair, still wearing oilskins and hard hat, and fell fast asleep.

The deep depression which had caused the storm drifted away towards Ireland, and although it was still blowing hard the following day, the wind moderated sufficiently to allow a mechanic to fly out from Yarmouth. There was not enough room on the helideck for another helicopter to land, and it hovered while the mechanic and his toolbox were lowered down on a winch rope. It took him less than fifteen minutes to discover that it was an electrical fault, not mechanical, and that we could take off. The small fault had cost a lot of time and money, but only a fool would have argued that the pilot had made anything but the right decision.

When I returned to the Hewett at the end of my two weeks off, both Vernon and Bill Macmillan were ashore. Bill's relief was J. C. Youngblood, a friendly, brisk and methodical man of medium build and middle age, always on the move, whose kindly face fooled many into making the mistake of believing that he was easy-going. Petroleum engineers did not get the responsible positions they had through being nice to people, and when J.C. was roused, men would keep well clear of him.

'That 'ole sonofabitch's meaner than a rattler!' declared Mitch Green ruefully, when J.C. had left the radio-room having made the point very forcibly that there was only one way a particular procedure was going to be carried out, and that was his way. I was startled to hear Mitch speak, and thought at first he was thinking out loud, and when he actually turned to me and spoke again, I began to wonder if I was hallucinating. In all the weeks I had been on the rig, I had never got more than a few incomprehensible grunts out of him. He had made it abundantly clear that, compared with Limey radio-operators, the alligators in the swamps of Louisiana were lovable creatures. Now here he was, actually speaking to me and smiling. I wondered why alligators had suddenly gone out of favour, and put it down to loneliness.

In many ways I had a privileged position, wedged between the bosses on the one hand and the men on the other. I knew almost everything that was going on aboard the rig, the conflicts, the resentments, the blunders, the successes and the personality clashes, and I had to tread carefully along a narrow path of diplomacy. To the tool-pushers and the Phillips men, I became someone they could confide in or let off steam to. Whatever Mitch's motives were, it was a pleasant change to sit and talk to him rather than have to suffer the silent antagonism that had filled the room whenever he entered. I still had great difficulty in tuning my ear to his soft, and at times almost inaudible, voice; but at least we were heading in the right direction. He knew everything there was to know about drilling but lacked Vernon's brash personality and verve.

The difficulty in recruiting roustabouts continued to be a problem. Although there were no more threats of strike action, few men stayed more than two weeks, and there were some who stepped out of the helicopter, stared incredulously at the noisy, mud-streaked, windswept metal island and hurriedly got back in again. Among those who did adjust to the hard working-conditions, there was considerable resentment that the Americans on the rig were earning about three times more than the British working on the same job. Mitch was howled down when he tried to explain to an angry crane-operator that Americans were paid an allowance for working overseas and that the crane-operator was earning twice as much as he would be earning ashore, plus free food. Mitch did, in fact, try to get a better deal for the British workers by asking Rimrock Tidelands to enquire about the possibility of paying their wages into a bank in the

Channel Islands to take advantage of the tax concession, but the Inland Revenue warned that anyone domiciled in the UK would still have to pay tax on any earned income brought into the country, so the idea was dropped.

Near the end of December, the rig Constellation helped the shareholders of the California-based company Atlantic Richfield to pay for their Christmas turkeys by striking gas about twenty miles north of the Hewett. Yarmouth buzzed with excitement at the likelihood of more prosperity coming to the town, but although service companies rushed in to pluck the golden goose, the sad fact was that few of them were capable of producing work to the very exacting standards demanded on the rigs.

On the Hewett, welding companies came and went like sparrows to a bird-table, and had they first made an effort to find out what was required of them, they would have saved themselves a wasted journey. Small engineering firms were often unable to read detailed drawings, and the proprietor of a catering company claimed that he worked out a tender for supplying food to a rig by eating as much as he could in one day. He lost the contract. Yarmouth businesses seemed to think that all they had to do was open their doors and money from the rigs would come pouring in. Rig managers dreaded anything going wrong on their rig between five o'clock on Friday evening and nine o'clock on Monday morning, for at weekends it was impossible to get anything or anyone, and there were times when I must have spent hundreds of pounds making radio-telephone calls, desperately trying to find a supplier who would stir himself to deliver a vital piece of equipment to the heliport.

It was difficult even during office hours to convey a sense of urgency. On one occasion, having called a company long established in the town and placed an order for urgent items worth thousands of pounds, I could hardly believe it when the sales director said that he was about to go on his annual holiday and would attend to it when he got back. Before I could say anything, Vernon's large hand reached over my shoulder and grabbed the handset.

'Listen here, you Limey sonofabitch,' he shouted angrily into the mouthpiece. 'Mah name's Vernon Andrews and ah'm the

toolpusher on this motherfucking rig, and I ain't gonna sit
around with mah finger up mah ass while you take a vacation.
You get that equipment on the chopper tomorrow or you ain't
got no contract.'

There was a strangled gasp at the other end of the phone, as if
the sales director had fainted with shock. When the helicopter
arrived the following afternoon, all the items were on board with
a compliments slip from the company addressed to Mr Andrews,
the 'tower splasher', thanking him for his esteemed order and
assuring him of their best attention at all times. I discovered later
that to get the items the company had had to send a van through
the night all the way to Aberdeen, and no doubt charged heavily
for it, but to an oil company or drilling contractor faced with
losing vast amounts of money if a rig was idle, instant action
was more important than the cost. It was something that the pil-
lars of the Yarmouth business community took a long time to
understand.

When I was at sea, I seldom got home for Christmas, and my young daughters had grown up believing that I was the Santa Claus who visited the house during the night to deliver their presents, then went back to his ship. Checking through the duty rotas for December, I was disappointed to discover that for my family this Christmas would be no different, except that this year Santa would be on a North Sea oil rig.

Tough though the American oilmen liked to appear, the hard veneer softened at Christmas time, and they were just as sentimental about it as everyone else. Phillips company men helped brawny drillers to string decorations across the mess-room and galley and make an artificial Christmas tree. The catering company sent out a real tree on the supply boat, but because of the fire hazard it was not allowed on board. The German boat crew fixed it on top of their wheelhouse and flouted all the rules for the identification of ships at sea by festooning it with coloured lights.

To celebrate the first Christmas on the rig, the chef produced a masterpiece of seasonal gastronomic extravagances, but the two essentials that make an oilman's life complete, an ample supply of booze and the warm company of women, were missing. In the rough, chauvinistic world of rigs of any nationality, women were strictly taboo, and while British rigs allowed two cans of beer a day per man and French rigs half a litre of wine with a meal, the Americans adopted the sober habit of their Navy and banned alcohol altogether. As with prohibition anywhere, ways and means were soon contrived to get round it, and after dark the strong tidal streams dispersed the tell-tale evidence that had once contained Scotland's water of life or the art of the brewer.

The tables were covered with fine white tablecloths, beautifully laid out with cutlery and table napkins, and in the centre of each table a miniature metal fairground carousel rotated slowly with the heat of a candle in its base. There was no swapping of presents, carol-singing or pulling Christmas crackers, but for each man there were two cans of a tasteless American non-alcoholic beverage, which helped to wash down a large helping of roast turkey cooked in a wine sauce. Drilling continued non-stop: for the men about to start the midday-to-midnight tour, dinner was breakfast, and for the men coming off the tour it was a welcome meal before they fell exhausted into their bunks.

Overcome with emotion in the season of friendship and goodwill to all men, one of the stewards touched the arm of an American driller while he was relaxing in the recreation-room and confessed that he was in love with him. With a startled yell, the driller shot to his feet and, grabbing the steward by the front of his shirt, lifted him into the air. 'Lookee heah, you bent sono-fabitch,' he bellowed, 'when ah gets me some pussy, ah likes to do it face to face.' Dropping the terrified man to the floor, he threw a chair at him and stamped out of the room.

On New Year's Eve I sat in the radio-room and listened to the traditional Hogmanay music of Jimmy Shand and his Scottish dance band on the BBC, then, with a few minutes to go to the end of the year, I pulled on a quilted jacket and climbed the steel ladder outside to the helicopter deck. It was a bright, starry night, and a hard frost glistened on the metal stanchions and stacks of drill pipe on the deck. Through the darkness, the anchor light of our standby ship gleamed reassuringly, and far away to the west the Norfolk coast glowed with light as if the whole population of Norfolk had gathered to dance and celebrate the great occasion.

It was a time for reflection and to philosophize. For me, the last two months had been the most memorable, the most re-warding and the most exciting time in my whole life. Whatever I disliked about the brash, uncouth, ill-mannered American oil-men, there was no escaping the fact that they were the hardest-working and most dedicated men I had ever met. Safe in plush offices in Houston, Oklahoma, Lafayette, Chicago, Los Angeles and New York, the oil-company moguls had no conception of the loyalty and dedication of the employees who slogged their guts out for them in the wild North Sea. I had seen toolpushers and Phillips men absolutely exhausted from days and nights without sleep, only to be assailed by memos or radio telephone calls from their shore bases or head offices. And for what? A large, tax-free salary in a job that was as secure as the life of a moth flitting around a candle. The blare of the standby boat's siren heralding the arrival of the New Year broke though my thoughts and, shivering with the cold, I returned to the warmth of the accommodation.

In his cabin opposite the radio-room, Vernon wearily kicked off his mud-stained overalls and climbed into his bunk. When he awoke, he would be hassled by Rimrock and hounded by Phillips Petroleum - tomorrow would be just another year!

The next morning, apart from hanging up new calendars and writing the date at the top of a new drilling-record sheet, there was nothing to indicate that it was New Year's Day. 'Happy New Year!' I called to Vernon when he shuffled into the radio-room clad in his thermal underwear and chewing a bacon sandwich. 'Yeah,' he drawled without enthusiasm, staring vacantly in the direction of Yarmouth - or perhaps America.

During the first week in January a vicious gale lashed the coasts of Britain. The Gulftide, a jackup rig newly built on the Clyde for a Gulf Oil-Rimrock Tidelands partnership, received a severe pounding in heavy seas as it was being towed to the North Sea, and the helideck sheared off and was lost. Two men on the rig were injured, but it was several days before the gale moderated enough to enable a doctor to reach them. Surging behind two large ocean-going tugs, the battered rig passed close to the Hewett on its way to the Wash for repairs before joining the search for gas off the Yorkshire coast.

My two weeks at home were spent climbing the peaceful, snow-covered mountains of the Lake District, but when I returned to the rig, life was anything but peaceful. I was hauled out of my bunk in the middle of the night by a foul-tempered Mitch wanting me to place an urgent call to Yarmouth. Something had gone wrong, and drilling was at a standstill. A grim-faced J. C. Youngblood came into the radio-room, and from his conversation with Mitch it soon became clear that the drill bit was stuck in a hole several thousands of feet down.

The holes being drilled from the Hewett were deviation holes, which meant that, instead of descending vertically beneath the rig to its maximum depth, the drill went vertically for perhaps a thousand feet and then, by means of a skilful technique, was made to curve away from the rig in a great arc. It was the intention of Phillips to drill several deviation holes radiating out in different directions and so tap the maximum area of the gas field from a central point. With the bit stuck several thousands of feet down in the hole, the danger was that, if too much torque was applied by the driller, it could snap the drill pipe. All the pipe above the break would then have to be pulled out and the lower section and the bit retrieved by a company which specialized in 'fishing'. It could be a lengthy process and the cost astronomical. While the drill pipe was still intact, there was a reasonable chance of freeing it by methods evolved over years of experience. There were no arguments or outward animosity. Mitch, J.C. and the driller discussed what was to be done calmly over innumerable cups of coffee and cigarettes, then went to the drill floor to try it out.

The air-waves were red hot with calls from the offices of Phillips and Rimrock in Yarmouth, demanding to know what was happening, but as the day wore on, the only certain information available was that somewhere far below the sea-bed there was a drill bit, well and truly stuck. Throughout the night the radio-room was like a siege headquarters, and I had to stay on watch making and receiving dozens of calls. Everyone seemed to want to get in on the act and, as it was daytime in the States, Rimrock's company president called several times for a progress report.

At first light, a helicopter flew in with drilling superintendents from Phillips and Rimrock on board, and they all spent the morning deep in discussion on the drill floor. A couple of attempts were made to rotate the pipe in the hole, and when that failed, the huge travelling block from which the pipe was suspended while drilling was in progress was hauled cautiously up the derrick to try to pull the bit clear, but it refused to budge. Calls were made to experts in the States for a second opinion, but all the brains, the experience and the concerted efforts of the experts and the men on the rig had no effect on the drill; it remained fast. It was generally agreed that most likely a large piece of rock had wedged the pipe or the bit, and movement in any direction only jammed it tighter, but the educated guesswork did little to sweeten the undisguised hostility between the Phillips representatives and Rimrock. For every day the drill was idle, the cost mounted like a ticking taxi-meter, and it was decided that the only course of action was to retrieve as much drill pipe as possible and plug the hole with cement. A new hole was spudded in, and the uneasy silence was broken as the drilling machinery roared into action.

The relationship between the Phillips company men and the toolpushers was very strained, and they hardly spoke to each other for days except to discuss essential business. When Bill MacMillan came out to the rig for his tour, I saw very little of him, and he would not discuss the incident, other than to say that it was one of the hazards of the game. Vernon was on the same crew-change but went about his work without his usual humour and enthusiasm, and the atmosphere in the radio-room was like a morgue.

The weather did little to lift morale, with one violent gale after another lashing the rig. The sea was so rough that the supply boats were not able to offload fuel and water; food and supplies were running low. On the rig the wind-speed indicator registered gusts of 70 mph, and further north it was even worse.

Early one morning, I was listening to a fleet of trawlers calling each other, and the language of the skippers was better than a barometer for giving an indication of the weather. At any time on the shipping frequencies the conversation of trawlermen was colourful, but as the weather deteriorated, the expletives became more numerous, and in really rough conditions they poured out in an unceasing stream. The barrage of obscenities from the skipper of a Hull trawler steaming up the North Sea bound for Iceland indicated that a gust of 80 mph had been reached on the ship's wind-recorder.

As he was talking, a distress message broke through on my emergency radio from the semi-submersible rig Sea Quest, owned by British Petroleum. The cables holding her nine anchors had parted in the storm, and the rig was drifting out of control a hundred miles east of Flamborough Head. British and Dutch tugs in the area immediately radioed the rig to see if they could assist, and a Shackleton aircraft of Coastal Command flew around close to the rig until help could reach it. Most of the crew were evacuated by helicopter, but the drama continued until shortly after midday, when the rig called Humber Radio cancelling its urgency message and reporting that, though still drifting, they were in no danger. High seas frustrated all attempts to take the Sea Quest in tow, and it drifted for nearly two days, with three ocean-going tugs from Hull in attendance, before they managed to save it from the ignominy of becoming a permanent tourist attraction on a Danish beach.

The tremendous storms created a crisis on the jackup rig Orion, working for the Gas Council/Amoco partnership, 150 miles off the Yorkshire coast. The radio-operator reported that the supply boats could not get to the rig, and work was virtually at a standstill through a shortage of essential supplies.

Life for the men working on the deck of the Hewett in conditions of gale-force winds and driving snow and sleet was extremely arduous. Though the drilling crew was protected from the full force of the gale by sheets of metal fastened round the base of the derrick, the biting wind blew over the top and through every chink in the steel grating, while snow and icy sleet poured from above, plastering the men from head to foot. For twelve hours out of every twenty-four, driller and roughnecks shivered and sweated on a cramped square platform, dodging whirling chains and trying to keep clear of the spinning rotary table and drill pipe while skidding about on a surface of mud and melting snow. The unrelenting snow affected the most resilient of the men, and morale was at an all-time low.

What was badly needed was a touch of humour to help dispel the depression, and it came in an unusual way. One of the most unpopular men on board was an enormous Texan from El Paso. He stood well over six feet four, with a body to match, and was the epitome of the clean-cut all-American boy, showering and changing his clothes at least twice a day, and immaculately turned out in well-pressed overalls, unbuttoned at the top to reveal six inches of hairy chest. He always wore a stetson hat and mirror sunglasses, indoors or out, and carried the tools of his trade in a leather holster slung like a six-gun on his right hip and fastened to his leg with a leather thong.

One of his tasks was to instruct the crew in fire drill, and one afternoon we assembled in the recreation room while, in his usual sneering tone, as if everyone but himself was an imbecile, he warned of the fire danger on the rig. Finally he grabbed a large foam fire-extinguisher in his outsized fist and held it aloft to demonstrate how to use it.

'This heah is a fire-extinguisher, and this heah is the tit that gives it lift-off. If there ain't a fire, the one thing you dumb Limeys ain't never got to do is hit the end of this sonofobitch.'

With that, he thumped the striker with his free hand. Instantly there was a loud 'swoosh' and he disappeared under a thick, clinging blanket of foam. As he stumbled round the room scraping it out of his eyes, the men were so helpless with laughter they could not move and clutched their stomachs, hardly able to breathe. The story spread round the rig within minutes and caused so much hilarity that even the Phillips man and the Rimrock toolpusher laughed about it together. It was just what was needed to break the spell of gloom, and life returned to normal.

The prosperity that the gas exploration and production brought to Yarmouth also created a number of social problems. It was natural that men cooped up on the rig for two weeks would want to relax and let their hair down when they 'hit the beach' with money in their pockets, and at times certain establishments were like Wild West saloons. Local girls were mesmerized by the mountain-sized, big-spending Americans, who wore stetsons and cowboy boots and called them 'ma'am'. Local lads, resentful that their women were being lured away by the boastful incomers, provoked bitter fights, and dance halls and pubs were often the scene of pitched battles, with chairs and bottles raining down on skulls and with floors littered with unconscious bodies, broken glass and pools of blood.

Once, fogbound in Yarmouth and unable to drive home, I was taken on a tour of the fleshpots by an American friend who had a penchant for young girls and strong drink - during his spell ashore he hit the town like a sex-craved bull dying of thirst. On our pub crawl the price of a whisky or a beer immediately doubled in many of the bars when they heard his accent, but undeterred he led the way from pub to pub and hotel to hotel, until my head was swimming and I longed for sleep. But for Gerry the night was young, and his eyes sparkled with anticipation. Standing in the middle of the road with the traffic shooting past him, he let out a wild 'Yahoo!' and threw his stetson into the air.

'C'mon on, Bob,' he yelled. 'We'll grab a bite to eat at a little 'ole place I know, then take in a movie.'

My brain was so numbed with whisky fumes that I could hardly focus my eyes, but I staggered along through the streets and up a flight of steps to the dining-room of a smart hotel. Gerry was obviously well known, and the manager himself brought the menus and fussed over us while Gerry ordered steaks and side salad for us both. When it arrived, the steak was as tough as old leather and accompanied by two slices of tomato, a piece of dried-up lettuce and a slice of cucumber, on a small plate. Any self-respecting rig catering manager would have thrown it over the side and fired the cook. During coffee, the manager sidled up and whispered in Gerry's ear, and a big grin spread over his face. Quickly downing the last of his coffee, he stood up. 'OK, let's go,' he whooped. 'The movie's gonna start!'

The manager led the way upstairs to a private room where a screen and a projector had been set up and easy chairs put in rows. A small, select group of cigar-smoking American oilmen, young Dutch and German sea-captains and one or two local businessmen in pin-striped suits were already occupying the available chairs, but the manager snapped his fingers and porters brought more chairs from another room. Stuffing Gerry's money into his pocket, the manager disappeared through a door and the lights went out.

The film started with three well-dressed, handsome lads and a girl sitting together in a lounge looking through a photograph album; then, for no apparent reason, one of the men suddenly took the album off the girl and started to undress her. Only minutes into the film, they were all naked and performing the most bizarre sexual gymnastics that left no orifice untouched and sprayed semen in all directions. It was the first blue movie I had ever seen and I felt slightly sick. Several other similar films were shown but there was not much variation on the theme, and to my relief the show ended.

But there was more to come! The speciality of the house, the *pièce de résistance,* took place in a room at the very top of the hotel, well guarded by heavies and accessible only to the favoured few who could afford it.

'Man, this is really somethin'!' promised Gerry, as we climbed more flights of stairs and stopped outside a door. 'This heah's a buddy o'mine,' he explained to the gorilla barring the door. Beady eyes looked me over for a second.

'OK,' he grunted, and opened the door to a lounge bar.

Except for a carpet, the room was empty of furniture, and the select group of fifteen or so lounged against the bar or the walls. A side door opened, and five naked girls trooped in and lay on their backs in a large circle, with splayed-out legs facing the centre. What followed was a sort of rampant rodeo as five men, having parted with £100 each, threw off their clothes and stood at the feet of the girls. At a signal from the hotel manager, they leapt into action to expend themselves, first on one girl, then onto the next in the circle, to the accompaniment of loud whoops and cries of 'Get to it, boy!' and 'Ride 'em, cowboy!' The girls lay passive and uninterested while the men humped and grunted on their bodies, but they did not have to wait long. Half drunk and lathered in sweat, the men were exhausted before they had finished the first round. With more loud whoops and 'yahoos' their friends cooled them off by pouring pints of beer over their limp appendages before stripping off their own clothes and lining up for the next session. In less than one hour, the drink, the heat and the vigorous exercise took their toll, and one by one they dropped out, gasping for breath. At a nod from the hotel manager the girls got to their feet and left the room.

Gerry was the last to give up and later, when we were having a drink in the hotel bar, his handsome features and massive physique won the affections of a waitress.

'Bob,' he grinned ruefully, 'Ah guess ah'll see ya back on the rig. Ah've got me some more tail!'

The men off the rigs were welcomed for their money, but at the same time looked upon as scum. If there was any trouble in the town, they were always blamed for it, and the police used the ruse of saying they were off the rigs when attempting to gain access to a pub or hotel suspected of serving drink after hours. Local newspapers had items headlined '£400 a month American fined for drinks offence', but there was never a headline proclaiming '£400 a month man keeps local economy alive by spending his money in hotels, shops, garages, coal merchants and milk suppliers, on the rates of his house and by driving a British car.'

Some local men who got jobs with the shore bases, or on rigs, entered into the spirit of things so enthusiastically that they became more American than the Americans. They adopted the oilman's dress, his speech and mannerisms, chewed gum or smoked large cigars and were so ludicrous they were dubbed 'Yarmouth Yanks' by the British rigmen. There was a story circulating about a Yarmouth Yank who went out to Texas to work on a rig and on his first day spied a couple of roughnecks sitting in the shade of a truck.

'Say, you guys, back home in England I was told that in Texas everyone toted a gun, but since I've bin here I ain't seen a single gun. All the guys on the rigs carry knives. Why is that?'

'Well, feller,' one of them drawled, 'in Texas, when a guy goes into the desert for a crap, he might get stung by a damn rattler and he has to cut the skin with his knife and suck out the poison.'

'You don't say!' breathed the Yarmouth Yank, wincing at the thought. 'Gee, you guys must be tough!' Then he looked puzzled. 'Say, if you were having a crap and the rattler stung you on the ass, how would you suck the poison out?'

'Waal,' the roughneck drawled slowly, 'that's when a man discovers who his friends are!'

'Mitch has quit,' said Vernon. I looked at him in amazement.
'You're joking!'

'No, I ain't. Only wish I was, but poor old Mitch has quit.'

'But why?' I asked. 'Surely they're not blaming him for
having to pull out of the hole and plug it?'

'Well no, I guess not,' declared Vernon, 'but in this game, if
something goes wrong and you're in charge, it's your ass that
gets chawed.'

It was sad news to be greeted with when I arrived back on the
rig after a spell ashore. Despite his odd ways, I had begun to like
Mitch and would miss him. Bill MacMillan was tight-lipped and
non-committal.

Mitch's replacement was a hefty, muscular man from Oklahoma called Ricky Reed. He had arms as thick as a ship's mast
and a voice that could be heard above the noisiest diesel engines
or the howl of a full gale. Almost as soon as he stepped aboard
on his first day, he went round the rig like a tornado, goading the
roustabouts into action from the helideck like General Custer
directing his troops against the Indians. Larry, a thin, laconic
farmer from Tennessee, who was helping me fit a new antenna
bracket on deck, listened to Ricky, then spat his wad of chewing-
tobacco over the side and remarked, 'Ah do believe we've been
fitted out with a noo foghorn.'

At the beginning of February we were plagued by gales, and
more crew-changes were carried out by the supply boats than by
helicopter. It meant having to transfer crew on and off the boat
with the personnel basket, and sometimes even the toughest of
men baulked at the prospect of being suspended by the crane
high above the sea.

'Are these ropes ever tested?' queried a young New Zealand
roustabout, nervously examining the well-worn personnel basket.

'Sure they are, son,' answered Vernon, with a twinkle in his eyes. 'Every time we send somebody down on it!'

When the gales ceased and the weather was settled, with the air warmed slightly, it attracted that other menace of the North Sea, thick fog.

In poor visibility it was always a nagging worry that a ship might collide with the rig, and although the foghorn had been repaired, it was not very powerful; in any case, it was well known that very few ships kept a look-out stationed in fog, preferring to rely on their radar sets. Radar was certainly a valuable aid, but for all their modern navigational gadgets ships still managed to crash into each other with alarming regularity. Within a few hours the fog was so thick that it was as if a massive blanket of grey cottonwool had been draped over the rig, cutting out daylight and reducing visibility to no more than three feet.

At 9 a.m. the skipper of our standby boat, the *Ross Delight,* reported that he could see on his radar what appeared to be a large ship approaching from the north, heading straight for the rig and about eight miles away. He had been trying to raise it on his radio without success and asked if I would have a go. I called Vernon and, tuning the radio to maritime calling and distress frequency, switched to full power.

'Hello the ship steering due south and approaching the drilling rig Hewett "A". Please take avoiding action. Our position is 52° 59' north, 09° 59' east.' I repeated it three times and waited for the *Ross Delight* to report.

'He's not moving!' said the skipper. I tried again. The skipper's voice broke the silence: 'He's six miles off and heading straight for you.'

'The sonofabitch!' swore Vernon. 'How long have we got?'

'Just over half an hour if he maintains his speed,' I said. Vernon made a quick decision.

'Give Bill MacMillan a call,' he rapped out to a roustabout, 'and tell him to get here pronto!'

'He's still heading for you,' said the skipper's voice.

Bill MacMillan came racing into the radio-room, rubbing sleep from his eyes. 'What in the hell's goin' on?' he demanded irritably.

'There's a goddam boat heading straight for us in the fog,' said Vernon quickly. 'The fishboat's seen him on his radar.'

'How far away?' asked Bill.

'About five miles now,' I said.

'Jesus!' swore Bill. 'We ain't got a lotta time to get off this mother. Better sound the alarm and get everyone up and dressed. Get that standby boat alongside damn quick!'

Vernon was reaching forward to press the general alarm button in a glass-fronted box on the wall when the skipper's voice spoke tersely:

'I think he's turning. Yes, he is!' There was a long pause . . . then the skipper's broad Yorkshire accent came over again:

'Aye, you're all right now. The daft sod must have remembered to switch his bloody radar on. He's turning to port and he'll clear you by about half a mile.'

'I guess I need a big cup of coffee,' said Bill, drawing hard on a cigarette.

Vernon gave a huge sigh of relief. 'Yeah, me too,' he said.

During February, a devastating epidemic of influenza swept through Britain, wiping out old people in their hundreds and almost bringing commerce, schools and industry to a standstill. To reduce the risk of its getting a hold on the rig, Phillips decided to have everyone on board inoculated with an anti-flu vaccine. Arrangements were made for a Yarmouth doctor to be flown out, but when it was revealed that he intended to take a female nurse with him, it caused more consternation in the headquarters of Phillips Petroleum than a sudden slump in the price of natural gas. No woman had ever set foot on a Phillips rig; it was unheard-of!

'Why can't you take a male nurse?' implored Phillips.

'Because I happen to employ a female one,' replied the doctor, 'and if you want me, you get her as well.'

The lights in Phillips' headquarters in America and London burned late into the night, and telexes flashed back and forth across the Atlantic as the cigar-chomping president and vice-presidents huddled together to discuss the crisis. World wars, stockmarket crashes, earthquakes and hurricanes they could deal with, but a woman on a rig: holy mackerel! But there was no option. The flu bug was sweeping through Britain, and a decision was needed quickly. It was reluctantly agreed, and a message was sent to the rig with instructions that the nurse could accompany the doctor to the Hewett, provided she went no further than the radio-room and the galley and that all pin-ups and girlie magazines should be removed from cabins. Why these should be removed when the nurse was to be confined to the galley and radio-room was not clear, but Ricky made sure that at least the more revealing and expressive pictures in the cabin art collections were hidden under pillows or in locker drawers.

On the morning of the medical team's visit, the rig buzzed with anticipation, and the effect on the men was staggering. Mud-stained hard hats had been polished until the aluminium shone, boots were scrubbed, coveralls were pressed and hair was slicked down with water. The stewards scrubbed the passageway and stairs until they shone; small bowls of artificial flowers miraculously appeared on the mess-room tables, and the cooks and galley staff were almost blinding in their starched white jackets and trousers. Bristows called to say that the helicopter with the doctor and nurse on board had taken off, and Ricky was enjoying a quiet coffee when Bill MacMillan said casually,

'Say, Ricky, what are we gonna do if this gal needs the john?' There was a choking sound as the coffee stuck in Ricky's throat.

'Jesus Christ!' he gasped. 'Nobody's thought of that. You're right. What the hell are we gonna do?'

Most of the toilets on the rig were rows of cubicles, and only toolpushers' and Petroleum engineers' cabins had *en suite* showers and toilets. Ricky looked pleadingly at Bill, but he stared out of the window and made no move to offer his facilities.

'Goddam it,' he said gloomily, 'I guess she'll have to use mine.'

When the helicopter landed and the engine was shut down, the doctor and nurse stepped onto the rig. It was an historic occasion which, had it been anywhere else, would have had full press and television coverage; but the only reception committee was a roustabout waiting to show them the way to the radio-room. Swarms of men clung to the guardrails and ladders, wolf-whistling as the petite figure swathed in a parka jacket with the hood pulled round her face descended the steps from the heli-deck. When she reached the door of the accommodation block and threw the hood back and smiled at them, they went wild with excitement.

Inside the radio-room she removed the parka and revealed a trim figure in a blue uniform and black stockings, her lovely auburn hair neatly tucked under a nurse's cap. She stood, vibrant and beautiful, whilst the doctor introduced them both and ex-plained what he wanted to do. All the time, Ricky stared at her open-mouthed. When she smiled sweetly at him and asked if she could wash her hands, he was completely flustered.

'Yessir, ma'am, you can wash your hands in mah john,' he said eagerly, then, realizing that she was British said, 'No, ah mean in mah, er, toilet john; hell, no, ah mean with the faucet in the john.' He realized he was getting himself in deeper and deeper. 'Aw, shit,' he said weakly, 'go through that theyar door marked toolpusher!'

Never was any gathering in the recreation-room so well at-tended. It was so packed that the doctor and nurse could hardly spread their materials on the table. Coolly ignoring the banter and pointed innuendo, the nurse held each man's arm whilst the doctor applied the anti-flu inoculation. Then she gently placed a strip of Elastoplast over the scratch.

'That Limey gal's the purtiest thing ah've ever seed,' sighed Herby, one of the derrickmen, who had a Mexican wife and ten children. 'I ain't gonna take that strip off mah arm till the day ah dies! When Juanita has another kid, ahm gonna call it "English Rose".'

Poor Herby was so captivated by the nurse that he overlooked the problems the name would create if his wife Juanita presented him with a boy.

Some of the men rejoined the queue for a second jab, to sa-vour the sensuous pleasure of the nurse's soft hands holding their arms, but the astute doctor had anticipated it and applied the inoculation to each man's left arm. Two Moroccan stewards almost had a pitched battle for the honour of taking the nurse a cup of coffee and biscuits, but the chef, adept at settling international disputes, made each take an end of the tray and carry it between them like altarboys bearing a gift to the Madonna.

The doctor and nurse were on board for two or three hours before the last of the forty crew had been dealt with, then, to a chorus of shouts and yells and waving of hats, they headed back to Yarmouth. It had been a welcome break in the dull routine, but the presidents and vice-presidents were right. The presence of an attractive woman on the rig was a disturbing influence: it drove home the isolation and magnified the harshness of the rough, tough world of offshore drilling.

In the remote lighthouses in wild locations off the British coast, the keepers spend their off-duty hours making models, woodcarving or studying, but for men on the rigs there was neither time, space nor peace and quiet to do anything but sleep, read magazines, watch a movie or sit in front of a TV if the rig was within range of a transmitting station. I spent what time I could get to myself designing my dream yacht and studying for a seagoing mate's certificate.

One evening, shortly after my transfer from the *Hercules* to the Hewett, my copy of the *Ship Captain's Medical Guide* attracted Vernon's attention, and he flipped idly through the pages.

'Say, Bob, d'you know first aid?' he asked.

'Well, I've got a Red Cross advanced first aid certificate,' I replied, 'but I'm no doctor.'

'Hell, I guess you are now,' he grinned. 'Ah've done promoted ya to rig medic.'

'Hey, sod that!' I protested. 'I've got enough to do. In any case, if there's a bad accident the first rule is to get a doctor.'

'OK, no problem,' said Vernon. 'If a man gets hurt bad, we'll ship him out on a chopper. All you've gotta do is fix him up temporary.'

Reluctantly I agreed, but little did I realize what it would mean. Men would appear at any time, day or night, wanting a wound dressed, a headache pill, cough medicine or advice on obscure pains. The pressure became so great that I had to organize regular 'surgery hours', and filled a small box with strips of Elastoplast and a few aspirins, with instructions that I was not to be disturbed unless it was a dire emergency.

The contents of the medical locker revealed a range of small bottles of medical magic with unpronounceable names, ampoules of morphine, sealed packets containing curved needles with a length of black thread attached, scissor-like instruments with serrated ends, a large bottle of saline solution with a long

plastic tube tipped with a fiendish hollow needle; a box of hypo-
dermic syringes, scalpels, surgeon's gloves, thermometers, a
tank for sterilizing instruments, and a whole range of bandages
and rolls of plaster, leg splints and dressings. Beneath a bench
were bottles of oxygen with face-masks, and a Neil Robertson
folding stretcher. Never believing I would ever have to use anything in the locker other than basic first-aid dressings, I found
the scalpels ideal for sharpening pencils, and one of the hypo-
dermic syringes perfect for squirting wood-worm killer into furniture at home. I sent my family doctor a list of all the drugs on
board, and it was returned with a neatly typed explanation of
what they should be used for and recommended doses.

I felt supremely confident - I could tackle any emergency -
and when Joe the chef rushed into the radio-room, gushing blood
like a stuck pig, I sat him down with all the panache of a Harley
Street surgeon. He had sliced deeply into his left hand with a
carving knife, between the thumb and first finger.

'I'll have to put a couple of stitches in it, Joe,' I said, pressing
a wad of lint against the wound in an attempt to stop his life's
blood ruining the polish on the radio-room floor. 'It's a nasty
cut!' Joe gulped and the colour of his face matched the starched
white of his chef's hat.

'Stitches?' he croaked. 'Are you sure you know what you're
doing?'

'Course I do,' I said. 'Tell you what. You just nip down to the
galley and tell your lads to get on with things, and by the time
you come back I'll have everything ready.'

Joe was still not convinced.

'Are you positive you know what you're doing? Have you
put stitches in before?'

'Many a time, Joe,' I said. 'Stop worrying. Away you go and
I'll see you in a few minutes.'

As soon as Joe was out of the room, I grabbed the *Ship Captain's Medical Guide* off the shelf and skimmed through it until I
found the instructions for stitching a wound. 'Clean the wound,'
I read, 'then prepare the length of silk thread attached to the sur-
gical needle, also the toothed forceps and scissors.' I was so en-

grossed in studying the diagrams showing how to push the needle into the skin that I did not notice that Joe had returned.

'Bloody hell!' he exploded, staring over my shoulder at the
book. 'You can't be that good if you have to look at a book to
learn how to do it!'

'Calm yourself, Joe. I'm just checking what size of needle I
need,' I said. 'Sit down and rest your hand on the edge of the
table.'

Drawing on a pair of thin rubber surgeon's gloves, I swabbed
the wound with antiseptic, then gripped a needle in the end of
the forceps.

'Are you ready, Joe?'

He was shaking like a leaf and drawing furiously on a cigarette.

'OK,' he wheezed.

I rammed the needle firmly into his skin, and the cigarette fell
from his mouth as he let out a piercing shriek. The chair fell over
with a crash as he scrambled to his feet and fled from the room,
trailing a long length of thread from the needle still stuck in his
hand. An hour later, when the helicopter arrived with the crew-
change, he leapt aboard with a towel wrapped round his fist.
Three weeks went by before he returned, healed and ready for
work.

'I've got a letter for you,' he said, breezing into the radio-
room and handing me an envelope. It was from the doctor in
charge of the casualty department at Great Yarmouth hospital.
Tearing it open, I unfolded the letter and read:

'Dear "Doctor", When you are next confronted with an injury
similar to that suffered by Mr J. Robinson, kindly throw your
bloody medical guide into the sea and use the enclosed, prior to
evacuating the patient to more knowledgeable care.'

Inside the envelope were three strips of sticking plaster. The
hospital was right to blast me for trying to be too clever, but it
did not alter the fact that in the absence of a trained medic on the
rig, I had been landed with the job of first-aid man, and there
would be times when helicopters would be prevented from flying and ships from reaching us. It worried me so much that during my spells ashore I enlisted the help of my family doctor, who

had been at sea in his youth and understood the problems. He arranged for me to go to a nearby hospital for several sessions, where a sympathetic surgeon taught me how to inject local anaesthetic and other drugs, stitch wounds and administer saline drips. I learnt a great deal from the surgeon, often beyond what a nurse would have been allowed to do in a hospital. He also made me attend an operation while he removed an appendix.

'You might never need to do this,' he boomed at me as I stood at the operating-table, gowned and masked, with half a dozen pretty nurses, 'but if your rig doctor can't get out because of the weather and tells you to perform an emergency appendectomy on a man, at least you'll know what he means, and not cut the poor bugger's balls off.'

With the problems of the first hole still fresh in their memories, the toolpushers and drilling crews were unusually tense, and tempers flared when anything threatened to disrupt progress. The unsettled weather in February meant that the sea was often rough, but when fuel and stores were needed, the difficult conditions were ignored and supply boats ordered to offload. The captains were highly skilled at manoeuvring alongside, but during one attempt thick nylon mooring ropes holding the *Smit Lloyd 6* snapped like pieces of cotton, and lines carrying fuel, water and cement were torn off the rig into the sea as the phenomenal power of the waves gripped the boat like a plastic toy and hurled it mercilessly against the rig. Roustabouts worked feverishly in biting winds, drenching rain and sleet to replace the lost lines, only to be defeated by the sudden onset of a wild gale which forced the captain to abandon any further attempts and pull away.

Frustrated and angry at the loss of the lines and at the possibility that drilling might come to a standstill because of the shortage of supplies, Vernon worked himself into an incredible fury, and anyone who crossed him found themselves on the next crew-change helicopter on a one-way trip to Yarmouth. He was worn down by hostile weather, irregular hours and rushed meals, and the strain of the responsibility of his job was etched into his face in deep furrows. The rig used up diesel oil and water at a tremendous rate; the drillers needed mud, drill pipe, casing and cement. With each hour stocks were getting lower, and as the gale raged a total shutdown loomed ever closer: if it happened, the company wolves would be howling for his blood! There would be no point in pleading that it was an 'Act of God' - the only god in the drilling business was money.

Not until the weather had moderated, the tanks were brimming with diesel, water and cement, and the deck stacked high with stocks of drill pipe and mud chemicals, did Vernon relax in his chair, heels resting on the edge of the desk, eyes half closed, a mug of coffee in one hand and a rolled cigarette in the other.

The morale of the toolpusher and drilling crew was also affected by the type of strata being drilled through. When it was fairly soft and drilling could continue with the same bit for several days, everyone was happy. If it was exceptionally hard rock and the drill bit wore out after only two hours drilling, it could take up to ten or twelve hours to pull the drill pipe out of the hole, one ninety-foot section at a time, and stack it in the derrick. Having changed the bit, it then needed another ten or twelve hours to run the lot back into the hole again.

The nature of the directional drilling on the Hewett necessitated specialist companies being brought in to monitor progress, using a system known as 'well logging'. By lowering a sophisticated array of instruments down the hole and interpreting instrument readings on the surface, a technician was able to form an accurate assessment of the different rock formations. A leading exponent of the well-logger's art was a French company called Schlumberger, which shipped out a portable instrument cabin to the rig in the charge of Louis-Jean, an excitable Parisian who chewed garlic at the breakfast table and chain-smoked Turkish cigarettes. One of Louis-Jean's tasks was to monitor the direction in which the drill bit turned, so that each hole was deviated in a different direction to finish at target depth, several thousand feet down and anything up to a mile away from the rig. A sensor lowered down the drill pipe sent back signals which gave the heading of the drill in degrees like a ship's compass.

Geologists employed by Phillips kept a watchful eye on the cuttings of rock being brought to the surface by the circulating mud. They were carefully separated on a vibrating screen and analysed under a microscope before being sent off in bags to Yarmouth under tight security. Geologists tended to be solitary people who never talked about their work and guarded their heaps of carboniferous compost like a mother hen protecting her chicks, but one jovial Dutchman took life less seriously than his colleagues. He was a keen yachtsman and an authority on Dutch sailing-barges, and we often sat and talked long into the night about sailing and boatbuilding. Sometimes he brought me plans of boats and a heap of sailing magazines out to the rig. Though they were printed in Dutch, it did not matter: the lines of a sailing vessel are, like music, an international language.

Compared with some of the giant platforms that have since risen out of the sea off northern Scotland, the Hewett was minute, and living-space was very cramped. I was fortunate to have a cabin which only on occasion did I have to share with visitors, but the men slept four to a cabin, with barely room to dress in comfort, and the only source of fresh air and natural light was a small porthole. What passed for an air-conditioning system was no more than warm air blasting out of a vent in the ceiling, and cabin interiors were uncomfortably hot and stuffy. There was an oversized cabin, euphemistically referred to as 'the recreation-room', in which the men gathered for a smoke or to watch TV if atmospheric conditions did not distort the picture. Twice a week a 16mm sound projector was unearthed to show movies which, when the sharp businessmen on the supply ships had negotiated a price, were augmented with the Continental capers of the blue movie industry.

The arrival of a mail sack twice a week was eagerly awaited, and the lucky ones would creep away for a few minutes solitude. There were men for whom there was never as much as a postcard, but each time the mail arrived they would rush to the letter rack, stare miserably at the empty boxes, then with a shrug of the shoulders turn away.

I spent most of my time in the radio-room, and it became a sort of refuge where men dropped by in the evening to sit and talk.

Larry, the Tennessee farmer, was a loner and normally kept very much to himself. The Americans called him 'the goofy boy', and big Tex, in his usual obnoxious way, sneered, 'That guy's so dumb he cain't walk and chew gum at the same time.' But I got to know him when he was sent to help me fix antenna brackets and move the heavy emergency batteries, and I liked him a lot. He showed me photographs of the small family farm at the foot of the Appalachian mountains, the timber-framed house with its veranda and rocking-chairs; of his mother, father, seven brothers and three sisters, his grandma and his old plough horse; but his eyes really glowed when he handed me a photograph of a smiling girl in a cotton dress seated on a tree stump with a dog on her lap.

'That's Rose-Ann,' he said proudly. 'We're fixing to get married when ah've saved a thousand dollars.'

There was always a letter in every mailbag from Rose-Ann, and Larry would bring it to the radio-room in the evening for me to read out. At first I found it embarrassing, until I realized that I was the only one on the rig he would talk to and confide in. It was not until a few weeks had gone by that I discovered to my amazement that he could hardly read or write.

'Ah didn't get no proper schoolin'. Mah folks were too poor,' he explained unhappily. 'Ma learnt us what she could; she learnt Pa too, 'cause he didn't have no schoolin' neither. Rose-Ann's had schoolin'. Her Pa owns the store.'

'Look, Larry,' I said. 'Some people get a start in life and some don't. There's plenty of people in Britain who can't read or write. It's not their fault; it's the fault of the system. Like your country, we live in a society where money and possessions mean more than people. What matters most to you is that Rose-Ann cares enough about you to want to marry you. She doesn't give a damn whether you can read or write, so forget it!'

I felt as if I had been preaching a sermon, but it was what he wanted to hear. While he wrote the words laboriously on a sheet of paper, I helped him compose a letter to Rose-Ann.

After the gales of early February there were a few days of settled, frosty weather, then the wind went to the south-west again and brought mild weather and heavy rain from the Atlantic.

In a heated argument over rates of pay, one of the crane-operators called Ricky 'a dumb Yankee bastard' had found himself sitting on the supply boat within the hour, heading for Yarmouth in search of a new job. He was replaced by Ron Gower, a tough Liverpudlian, equally capable of using descriptive language but rather better at choosing his timing. One of his team of roustabouts was Charlie, a man in his forties who stood out among the assortment of young Australians and New Zealanders, though it was more for his greater capacity for work than the difference in his age. He said little about himself other than that he was from London and that he and his wife had once managed a pub but had given it up.

One evening Charlie came to ask if I kept a drug called Prednisolone in the medical locker. I said no and asked what he wanted it for. He said he suffered from bad headaches and he had run out, but as he was going ashore the following day he would get a prescription from his doctor. I gave him a few aspirins and he went to his bunk. Curious to find out what Predisolone was, I asked a doctor friend when I went ashore, and was staggered to learn that it was generally used as a pain-killer in the treatment of cancer. I ordered a small bottle of the tablets from the wholesale chemists who supplied the rig, and secured them in the medical locker. A few weeks later, Vernon told me to go and look at a man lying in his bunk unable to get up for work, and when I went to the cabin I found it was Charlie, lathered in sweat and writhing in agony. He had used up his supply of tablets. I gave him the Predisolone, left him to sleep and told Vernon that he had diarrhoea.

'I think you'd better tell me what you're really suffering from,' I said, when Charlie came to see me later that evening.

He told me an incredible story about how he had been forced to give up managing the pub when his wife had developed multiple sclerosis and become confined to a wheelchair. Shortly afterwards it was discovered that he had stomach cancer, and no one would give him a job. He and his wife were so desperate for money that he went to Yarmouth, told a few untruths and was

taken on as a roustabout and sent to the Hewett. When he finished speaking, he slumped into a chair and buried his head in
his hands.

'I suppose you'll have to report me now,' he said despair-
ingly.

I had been taken in by so many hardship stories when I first
joined the rig that I had become hardened to them, but there was
something in Charlie's voice that rang true, and it put me in an
awkward dilemma.

'It's not for me to report you,' I said. 'All I'm required to do
is to tell Vernon whether or not you're able to work. If you say
you are, that's good enough for me.'

He looked up in surprise. 'Yes, I'm OK now. Can I keep the
tablets?' I nodded. He turned to the door, then stopped and
shook my hand. 'Thanks a lot,' he said quietly. 'You don't know
what it means to me.'

Having done my bit for humanity, I wondered if the company
would be as compassionate with me if the truth got out.

Another regular evening visitor to the radio-room was Scotch
Jimmy, a strange, wild-eyed ex-docker from Glasgow who was
fighting a drink problem. When he struck a match to light a cigarette, his hand shook so violently that it took several attempts
before he managed to get the flame and the end of the cigarette
together. He had been in gaol for having an incestuous relationship with his teenaged daughter, which his wife had happily
condoned.

'She said she didna care who I had it away with, so long as I
kept off her,' Jimmy said, unburdening his soul one evening.
'Och, it was an awful thing to do, but when a man gets the evil
drink inside him, he's in the grip of the devil himself!' He stared
at the ceiling, and suddenly his voice rose to a crescendo: 'I will
give thanks to thee, O Lord, for thou wast angry with me. Thy
wrath is turned away and thou hast comforted me!'

There was an awkward silence as he continued to stare at the
ceiling as though expecting a reply, but when nothing came, he
turned to me. 'Isaiah, Chapter 12,' he said, almost apologeti-
cally. Then with a shaking hand he lit a cigarette and inhaled
deeply. While he was in gaol he had 'seen the light', and his

knowledge of the Bible was astonishing, but there was one weakness of the flesh from which he would never be cured. The voice that gave fervent thanks to the Lord was well-lubricated with 'Nectar of Old Nick.'

During February a trawler triggered off a scare that a gas wellhead off the Yorkshire coast had blown out. The skipper reported that the sea was bubbling around the buoys marking the spot where the BP rig Sea Quest had been blown off location in the big January gale. Waterspouts were being thrown high into the air by the leaking gas, and the area was widening rapidly. The jackup rig Constellation was towed out to drill a hole near the well to relieve the pressure while the wellhead was repaired, and all shipping was warned to keep well away.

March lived up to its reputation and came in like a lion, with one fierce gale after another screaming down the whole length of the North Sea, piling up mountainous waves. For hour after hour the rig shuddered under the weight of thousands of tons of foaming water crashing against the legs with a sickening thud that woke the men who were asleep in their bunks and flung pans to the floor in the galley. The *Smit Lloyd 6* was on duty standing by the rig, but the captain was far too busy with his own survival to be much help to us. From the radio-room window I watched horrified as the boat lifted on a wave so steep that the mast was horizontal. It did not seem possible that it could do anything other than fall backwards like a man losing his grip on a rock face, but at the very last moment it teetered on the brink, then plunged down the back of the wave with both propellers spinning out of the water. Had a wave like that caught the helmsman unawares after dark, only a few pieces of floating wreckage would have been left to tell the tale.

In such conditions the rig was completely isolated, and the morale of the men plummeted when helicopters could not fly and crew-changes were postponed. Every extra day on the rig meant a day less spent ashore, and there was no compensation or days off in lieu. In their anger and frustration men became careless, and it invariably led to a spate of accidents.

The howling of the wind and booming of the heavy seas against the rig kept me awake one night, and I was reading in my bunk when my cabin door was thrown open and a roustabout shouted that a man had been hurt and was lying on the drill floor. Pulling on my clothes, I stumbled out onto the deck.

The force of the wind almost bowled me over. The air was thick with spray torn off the sea by the fury of the gale, and companionway ladders, stacks of drill pipe, guardrails and floodlights were encrusted with salt. I staggered across to the drill works and found the man lying unconscious on the deck. Above the roar of the gale and the noise of machinery, the driller yelled that he had been struck by a whirling chain while they were pulling the pipe out of the hole. With the aid of the folding stretcher, the drilling crew carried him to the radio-room, dumped him unceremoniously on the floor and went back to work.

Blood was oozing from a bad wound on his left thigh, and I was about to cut through his trouser leg with scissors when he suddenly regained consciousness and refused to let me near him.

'Look,' I said. 'Either I remove your trousers or you bleed to death. It's up to you.' He mumbled something about not wanting anyone to know.

'Know what?' I asked, but he just moaned and lapsed into unconsciousness.

I quickly unfastened his waist belt and pulled his trousers down to his knees; then I stopped, hardly able to believe my eyes. His right leg had been amputated above the knee, and he was wearing an artificial limb. The wound on his left leg was bleeding badly, and I closed it with five stitches, covered it with a dressing and gave him an anti-tetanus jab.

One of the men who helped to carry him in had said that when the chain hit his leg he had slipped and banged his head on the steel grating. I examined him closely but apart from a nasty bruise on his forehead there was no other head injury, and having dressed him in my pyjama trousers and socks to cover the artificial limb, I summoned help and had him deposited in the spare bunk in my cabin. I stayed awake for a while to make sure that he was not in pain, and as I listened to his rhythmic breathing I could not help having the utmost admiration for his courage. He had worked on rigs for over three years and gone from roustabout to roughneck without anyone's being aware of his disability. What on earth was I going to write in the accident report?

When I woke him with a cup of tea the following morning, that was the first question he asked me.

'All I know about the incident,' I said, 'is that I discovered you unconscious, having sustained a blow to the head and a laceration to the left thigh. I inserted five stitches in the wound and dressed it, gave you an anti-tetanus jab, then had you removed to a bunk with suspected concussion. Is there anything I've missed?'

He smiled and gripped my arm, then with a long sigh sank under the blankets and fell asleep. The cup of tea lay untouched, and I emptied it out of the porthole. Two days later, the crew-change helicopter managed to reach the rig and he went away, having completely recovered.

'Don't forget to call at your doctor's and have the stitches removed,' I said, as he passed me on his way to the helideck.

While he was at home he was offered a job as a driller with a company drilling offshore in Nigerian waters, and I never saw him again. Several months later a magnificent book about boat-building arrived in the post, with a brief note that read: 'With grateful thanks from Pegleg.'

The crane barge *Hercules* prepares to lift the accommodation
block, engine-room and drillworks of the Hewett A

The engine-room of the Hewett A is hoisted onto the platform

The helicopter pilot's view of the Hewett A showing the landing platform (*left*), the accommodation block (*beneath it and centre*) and the main drilling-platform (*right*) with the derrick rising out of it

Heavy seas halt pipe-laying

The drilling-platform of the Hewett A ready to receive the derrick

The author in the radio room. Spartan though life was, we did have curtains!

Dinner-time — American style. Good food and plenty of it!

The drill floor − preparing to stack drill pipe in the derrick

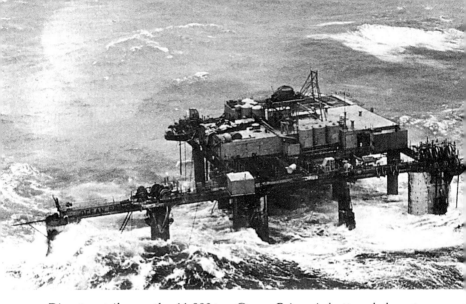

Disaster strikes as the 11,000 ton Ocean Prince is battered almost beyond recognition by waves driven by a seventy m.p.h. wind

The supply boat *Hector Gannet* (*pictured here*), which had attempted to pick up survivors from the Hewett A, sinks after striking the rig. Ploughing to the rescue in the background is the trawler *Boston Hornet*

Red Adair, the Texan expert in oil and gas fires, at a press
conference after the blow-out

On 5 March 1968 the BBC shipping forecast warned of further severe gales in the North Sea, with winds reaching hurricane force at times. Work on the Hewett was brought to a halt, and only essential jobs were carried out on deck. Ships were in difficulties over a wide area, and rescue services were hard pressed answering calls for assistance. A Dutch helicopter with twenty men on board took off from the Netherlands to fly out to a rig and vanished without trace. Near the Danish coast the crew of an oil rig watched helplessly as their supply boat was engulfed by enormous seas and driven under. I listened in to the dramas as the brave men in the lifeboat services of Denmark and the Netherlands battled against the seas in a vain hope that they might find survivors; but, as their tiny craft climbed to the tops of the watery mountains rising in front of them, there was only a terrifying expanse of mile after mile of angry white waves in which no human being could live for more than a few seconds.

To the north of the Hewett the first semi-submersible rig to be built in Britain, the gigantic 10,000-ton Ocean Prince with a crew of forty-five, was drilling for Burmah Oil on the Dogger Bank, a hundred miles east of the Humber. It was accompanied by the converted stern trawler *Hector Gannet* - the boat that had assisted Mr Cap, the first rig to drill in the North Sea.

In the shallow water around the bank the rig was able to rest on the sea-bed instead of floating, but the scouring action of the waves on the sea-bed caused the rig to settle deeper than was normal on its pontoons, and as the weather worsened, things started to go wrong. At about two o'clock on the morning of Wednesday 6 March, a large wave struck the rig with such force that it threw men working in the engine-room to the floor. Drilling was stopped and efforts were made to pump the water ballast out of the pontoons to float the rig off, but it was too late. The rig was stuck fast in the sand at the mercy of the hurricane. Huge seas smashed into it, sending the stacks of casing and drill pipe rolling across the deck, threatening to annihilate anyone who got in the way.

An hour later cracks began to appear in the deck, and the main truss holding the sides of the rig together and supporting the 140-foot-high derrick began to work loose. It was obvious that the giant rig was breaking up and the derrick was about to collapse. A distress message was put out by the radio-operator asking for immediate assistance, and all the crew put on life-jackets and huddled together away from the swaying derrick.

At seven o'clock the derrick finally collapsed into the sea, taking with it two cranes, the radio shack and a large section of deck. What little deck remained was supported precariously on the main column, which creaked and groaned in the maelstrom of foam that poured over the rig lashed by winds gusting to 90 mph. The air was filled with spray that soaked the men crouching by the helideck, but they hardly noticed it. All around them the latest in multi-million-pound computer-designed technology, as large as a football pitch, was being pounded into a heap of scrap metal by the rampaging forces of nature. The power of the seas bent and twisted metal girders like fuse wire, jammed metal doors and forced windows out of their frames. More and more of the rig collapsed into the sea, and as the men watched, they knew that soon it would be their turn. The standby boat, *Hector Gannet*, was riding out the storm but, though the crew could see the men on the rig huddled together, it was too dangerous to go to their aid. It seemed as if nothing could save them.

But help was on its way. Odeco, the rig's owners, had received a call from the rig before it started to break up and had alerted Bristow helicopters at Scarborough to evacuate the rig as a precaution. Captain Bob Balls, the pilot who had been at the Sea Gem tragedy in 1965, was the duty pilot, and at 5.45 a.m. he took off alone in a sixteen-seater Wessex for the long flight to the rig. It was pitch dark, and over the sea he felt the full force of the gale as rain and sleet battered against his windscreen. Dawn was breaking as he approached the Ocean Prince's location, and below him the sea was a raging mass of white foam. When he saw the black mass of the rig and noticed that the derrick was missing and the deck seemed to be in two sections, he realized that it was more than a routine evacuation: it was a real emergency, and time was running out for the tiny figures clustered by the helideck.

Landing on the shaking rig, he indicated by signs that he could take no more than eighteen, and as soon as the helicopter was full he took off and headed for the rig Constellation, which was about twenty miles to the south. Establishing contact with the cabin passengers through an intercom, he was alarmed to discover that there were nineteen men on board, which was over the weight limit for safe flying, but there was no turning back. As soon as he had dropped the men on the Constellation, he set off again for the Ocean Prince.

Bob had worked out that he had enough fuel for another round trip, but by now the wind was way beyond the helicopter's normal operating limit and he was heading into it. For the men left on the Ocean Prince, it was agonizing watching the nineteen flying off to safety, and waiting for the helicopter to return seemed an eternity. Back at the Ocean Prince, Bob lifted off another eighteen, leaving eight to sit it out on what was left of the once proud rig.

The remaining eight were particularly brave to elect to step back and let others go first, and as the rig twisted and jerked under their feet in the final throes of death and the seas reached for them, the strain must have been almost unbearable. Their suspense was made worse by the helicopter having to spend precious time refuelling on the Constellation, but it was a job that could not be rushed. To get water in the fuel would have been disastrous, and the fuel supply had to be tested and filtered before it could be pumped into the helicopter's tank.

At last Bob was able to get airborne again, fully aware that when he arrived at the Ocean Prince there might be nothing left but a few pieces of metal sticking out of the water marking the grave of the eight men he had been forced to leave behind. Peering through the wind-driven sleet, it was difficult to distinguish between the sea and the sky, and relying on his altimeter to maintain a safe height he headed into the blizzard. Arriving at the rig's location, his heart skipped a beat. There was nothing but sea! Then he saw it . . . a tiny black speck, like a pinhead in the white foam, was all that remained of 10,000 tons of welded steel. The helideck was already tilting at an angle as the remains of the rig began to sink deeper in the water, and with the eight men safely aboard, Bob lifted his helicopter away for the last time. The ordeal was over, but not a moment too soon! Shortly afterwards, the pilot of an aircraft flying over the scene reported that there was no sign of the rig and he could not pick it up on his radar.

For his bravery, Captain Bob Balls was awarded the MBE.

On the same day, not far from the Ocean Prince location, another semi-submersible rig, Ocean Viking, drilling for Phillips Petroleum, radioed a warning that waves estimated to be at least sixty feet high were smashing into the rig and that they had been forced to cut their anchor cables and drift with the storm. All the men were still on board.

The Hewett suffered very little damage from the hurricane, except for the wind anemometer's shearing off its mount and falling into the sea, and my main transmitter antenna wire being blown off the top of the derrick; but a day after the Ocean Prince tragedy I was involved in a little drama of my own.

A trawler in the vicinity radioed to say that the ship's bell had broken loose in heavy seas and struck one of the crew on the head; he needed urgent medical treatment - could we assist?

Vernon was the toolpusher on board, and I asked whether it would be possible to lift the man off with the personnel basket. He said it would be risky but he would try. I called the trawler and said if he could get alongside with the man ready to be lifted into the basket, the crane-operator would be waiting. There was a nasty sea running, but for once providence was on our side, and by a combination of skilful seamanship, quick action by Ron Gower, the crane-operator, and a timely smooth patch in the seas, the man was hoisted onto the rig and carried into the radio-room.

He was unconscious and covered in blood and had the most appalling gash across the top of his skull. Bill MacMillan radioed the Phillips base for a helicopter while I set about cleaning the man up and stopping the blood flowing from his head. I had to cut off all the hair around the five-inch gash in his skull and clean it as carefully as possible before applying thin strips of plaster to stop the wound opening further. He regained consciousness and was obviously in severe pain, but my first-aid training had warned against giving morphine for head injuries. I put a call through to the hospital in Yarmouth, and a doctor instructed me to inject the man with Fortral, which by a stroke of luck I had in the medical locker.

When the helicopter arrived, the wind was gusting to gale force, but the pilot brought it in magnificently and within seconds was airborne, with the injured man still clad in his oilskin suit and reeking of fish. He was in hospital for many weeks with a fractured skull, but fishermen are made of tough material, and as soon as he was pronounced fit, he went back to sea on the same trawler, all thanks to that most versatile of flying-machines, the helicopter, and the daring of the pilot.

The wire antenna, shearing off at the top of the derrick, put my main transmitter out of action, but secretly I was very pleased about it. I had been looking for a good excuse to go up the derrick, but the toolpushers were reluctant to allow anyone but the drilling crew to climb the 130-foot vertical metal ladder.

'It's a mighty long way to the deck when you're at the top,' warned Vernon, after I had persuaded him that it was essential I check the antenna mounting bracket to see if the storm had damaged it. 'Ah hope you gotta good head for heights. Make damn sure ya wear the safety harness.'

Stuffing pliers, a short coil of wire and a camera into my windproof anorak, I tied one end of a thin rope round my waist and the other to the end of the broken antenna wire, so that I would be able to haul it up after me, and stood at the foot of the derrick. The canvas safety harness was attached to a wire rope which went over a pulley at the top of the derrick and was then secured to a balance weight. I fastened the harness round my chest and released a catch holding the wire. The purpose of the balance weight was to arrest a fall should the climber slip off the ladder, and it certainly worked - so effectively, in fact, that as soon as I released the wire and the weight took over I shot up the first fifty feet of ladder like a sack of corn being hoisted to the top floor of a flour mill. I grabbed the steel rungs and hung on until I got the feel of the weight, and the rest of the climb was effortless. It felt as if someone was sitting at the top of the derrick heaving me up, and I reached the derrickman's platform at ninety feet with my boots hardly touching the ladder.

Below me the deck had shrunk to the size of a postage stamp, and tiny figures were working with a toy crane. There was only a slight breeze, but all around the rig, stretching to a dark horizon, the sea was white and angry in the aftermath of the storm. I looked up at the forty feet of ladder still to climb, and for the first time began to feel apprehensive. A keen mountaineer, I had no fear of heights, but unlike a solid rock face the derrick was a man-made structure which was being constantly shaken by the machinery below it, and it gave the impression that the whole lot would topple into the sea at any moment. The fear of being ridiculed for turning back has probably spurred on many a man to glory and a medal in wartime, and it was the same fear which helped me to the top of the derrick. Gritting my teeth, I stepped off the safety of the derrick-man's monkeyboard and climbed slowly to the last rung, 130 feet above the deck and nearly 200 feet above the sea.

In front of me was the huge crown block supporting the cables of the travelling block, which in turn took the weight of thousands of feet of drilling pipe in the hole. It did not seem possible that such a flimsy structure could withstand the enormous strain. Looking down at the deck was nerve-racking, but the view around was fantastic. It was like sitting on the top of an electricity pylon surrounded by sea, and though there was a lot of cloud about and the horizons were hazy, Cromer Head on the coast stood out well, and the sea seemed to be alive with ships heading in all points of the compass.

The antenna wire had snapped close to a porcelain insulator, and it took only a matter of minutes to haul the wire up on the end of the rope and secure it. Quickly taking a few photographs, I climbed back to the safety of the deck and the radio-room. I had achieved my ambition to climb the derrick - but once was quite enough!

I spent the rest of March at home among the peace of the mountains, glad to be many miles away from the intensity of life on the rig and the capricious North Sea weather.

When I returned to the rig, the weather had considerably improved, drilling was on schedule and everyone was contented. Larry was over the moon about receiving letters in the mail from Rose-Ann, and he brought them for me to read. She said how much she missed him and how she was looking forward to the wedding. Her Pa was busy making plenty of liquor for the reception and hiding it in the hills, and her Ma was going to make a wedding dress from a length of white silk that a travelling salesman had let her have for a few dollars. Larry savoured every word, and there was no happier man on the rig.

Jimmy had been in trouble with the law again and summoned to appear before a Glasgow court, charged with indecent exposure. Having spent the first morning of his shore leave helping to double the sales of the local whisky-supplier, he had preached a fiery sermon in Kelvin Grove Park about the evils of drink and the temptations of the flesh; then, according to his version of the story, he went behind some bushes for a pee and was wrongfully arrested. The police, however, pointed out that not only did he continue to shake his appendage considerably longer than was necessary to remove the last few drops but he chose to do it in front of a party of little girls on their way home from school.

It was only the defending lawyer's masterful pettifoggery that saved Jimmy from a month or two sewing mail bags, when he persuaded the gullible judge that it was not Jimmy's morals that were suspect but the condition of his bladder, and having a medical condition he had to empty it. He had cleverly avoided mentioning that the only medical condition Jimmy had been treated for was alcoholism, and went on to make an impassioned plea about how his client had reached the point of no return when the little girls unfortunately walked by. The case was dismissed, and the arresting police officer nearly burst a blood vessel with anger and frustration.

Charlie, the other regular evening visitor to the radio-room, looked pale and washed out but confided that the pain in his stomach, if not getting any better, was at least no worse, and he was able to contain it with his tablets. He was more concerned about his disabled wife than his own condition. Her debilitating disease was beginning to affect the co-ordination of her hands, and he was worried that a next-door neighbour who looked after her while he was on the rig might soon find it too much. Outwardly he was a cheerful and hardworking man, and no one was ever aware of the dreadful problems he had.

Drilling had progressed without meeting any snags, and the final string of casing had been cemented in at target depth. Louis-Jean, the garlic-chewing Frenchman, arrived to carry out a very clever technique by which the gas could be released to flow to the surface. The tubular steel casing tapered from twenty inches diameter at the rig to seven inches at target depth several thousand feet below, and Louis-Jean lowered a special gun inside it to the bottom of the hole. When activated electrically from the surface, the gun fired several bullets which perforated the sides of the casing and allowed the gas to flow to a control valve on the rig at a pressure of up to 5,000 pounds per square inch. It was a very delicate operation, and all the radio transmitters had to be shut down to prevent antenna radiation setting off the charges prematurely.

The normally affable Bill MacMillan considered the eating-habits of the French to be the lowest form of human life, and whenever Louis-Jean was in the same room he always made a point of opening all the windows.

'Ah reckon that froggy sonofabitch has more chance of setting the gun off when he farts garlic fumes,' he commented drily, after Louis-Jean had looked in to check that all the sets were switched off.

It took several hours to complete the operation, but there was one unmistakable sign that it was over! The sudden departure of Bill MacMillan in one direction and Vernon in another heralded the arrival of Louis-Jean crunching a celebratory chunk of garlic.

'I haf done ze perforation,' he announced solemnly. 'You may continue with ze radio.'

Leaving the door and windows open, I strolled round the deck for five minutes until the air in the radio-room had thinned to a pollution level more acceptable to Anglo-Saxon lungs.

With the improvement in the weather and the drilling programme running smoothly, morale was high, and the only black
cloud was the continuing war in far-away Vietnam. Almost
every American on board had a relative serving in the Army or
the Air Force, and as the newspapers and TV news reported the
carnage with every gory detail, even loud-mouthed Tex gave up
boasting that, 'Our boys will knock the shit outa the Cong.'

The flag-waving, chest-heaving politicians in Washington
waged their campaign unaware that the mental anguish and psy-
chiatric problems proliferated by the war were reaching way
beyond the jungles of that little-known country on the edge of
the South China Sea. If the news from Vietnam was particularly
gloomy, the atmosphere on the rig was tense and led to heated
arguments and sometimes fist fights between British and Americans.

There was an occasion when the crew-change helicopter was
delayed by a persistent morning fog and the men due to go
ashore were sitting patiently in the recreation-room, browsing
through newspapers or playing cards. Twice the helicopter took
off from Yarmouth but over the sea encountered thick fog and
returned to base. When I broke the news to the men, Tex kicked
a chair over in temper and flung his holdall to the floor.

'Goddam chicken-livered Limey pilots!' he sneered. 'Our
boys in Vietnam fly in anythin'.'

There was an awkward silence, then Yorkie, a little assistant
mechanic from Hull who even in high-heeled boots could never
reach five feet tall, looked up from the newspaper he was reading.

'Maybe that's why your boys bomb their own bleeding
troops,' he said angrily. 'Here, read that!'

He flung the paper at Tex, and it fell onto a table, its vulgar
headline spread out in bold black type: 'Yanks Bomb own
Troops in Vietnam.' Tex went beserk and, grabbing Yorkie by
the lapels of his coat, lifted him high into the air and shook him
like a doll.

'Don't give me no shit, you Limey sonofabitch!' he yelled.
'You ain't nothing but a goddam North Sea nigger anyways.'

The taunt stung Yorkie, and though he was no match for Tex in a fight, he did the next best thing. Twisting out of his coat, he kicked the big Texan in the balls. Tex doubled up in agony, but as he fell to the floor he managed to get one of his enormous hands round Yorkie's throat and would have squeezed the life out of him if the rest of the men in the room had not dragged the two apart. Tex was crazed with anger, and as soon as he was on his feet, he picked up a chair and attacked Yorkie again. It took the combined strength of three hefty men to restrain him until he had quietened down, and Yorkie took the opportunity to beat a hasty retreat and hide.

'What in hell was that ruckus about?' asked Vernon, looking up from the drilling-record book when I got back to the radio-room.

'Oh, it was only Tex getting a bit excited,' I replied. 'Nothing serious.'

'Huh,' snorted Vernon, 'that guy's plum loco.'

It was the frequent use of the word 'nigger' and the uncon-cealed hatred of black people that I found most distasteful and disturbing about the Americans from the southern states. In the search for international tolerance and understanding, there are many things which, though anathema to our own way of life, we have to come to terms with and respect as part of another coun-try's heritage and tradition, but man's inhumanity to man can never be one of them. On the rig, there was a man from Alabama who openly boasted that he had sued a hotel when he discovered that a Negro had slept in the room the previous night.

When Yarmouth employment office sent a black man out to the rig to work as a roustabout, there was almost a riot when some of the more bigoted Americans saw him on the supply boat, and they refused to allow him to be lifted onto the rig. The unfortunate man suffered the most degrading humiliation as he stood bewildered on the deck of the boat while obscenities and abuse were hurled at him from the rig. The strange paradox was that the same people would work in harmony with all races until the colour of a man's skin went beyond dusky brown to jet black, whereupon they lost all reason.

Early in the first week of April, when the world was shaken by the news that the black Civil Rights leader Martin Luther King had been shot and killed in Memphis, Tennessee, I was sickened by the hoots of laughter and the derisive comments from Americans for whom I had the greatest respect and feelings of friendship. It was difficult to comprehend that honest, hard-working men, who had families and loved-ones of their own, could talk gleefully about dancing on the grave of a man who had won the Nobel Peace Prize and who had devoted his life to campaigning for a better deal for the underprivileged in American society.

The North Sea was a popular fishing ground for trawlers of many different nations, and during a spell of settled weather the seas around the Hewett would be alive with German, British, Dutch, French, Belgian, Danish and sometimes even Norwegian and Spanish boats, all competing for the dwindling survivors of the once vast shoals of fish to be found on the shallow banks. Sometimes a boat would come alongside and barter for a frozen turkey or a leg of beef in exchange for a basket of fresh cod or plaice or occasionally the most delicious sole, on which the chefs demonstrated their gastronomic skills and produced a meal fit to tease the palate of the most discerning gourmet.

Towards the middle of April a large trawler appeared on the scene flying the hammer-and-sickle emblem of the USSR, but its mission had nothing to do with catching fish. The KGB had obviously decided that a few pretty pictures of British gas rigs would look very nice on Kremlin office calendars, so to give the official Party photographers a welcome break from snapping the jovial line-up of Soviet top brass viewing the May Day Parade, they were sent on a mini-cruise in the North Sea.

For an hour the trawler steamed round and round the rig, and a fur-clad figure perched on the top of the wheelhouse rather fancied himself as Russia's answer to Sam Goldwyn. Bawling instructions through a megaphone to the photographers on the deck, he rushed to squint through the viewfinder of a movie camera held by an assistant before shouting more instructions and dashing back for another squint. Between them, they must have shot hundreds of yards of film, but it was doubtful if any of the pictures would have been pronounced suitable for office calendars or 'Having a lovely time, wish you were here' picture postcards for sale in Siberian labour camps. Each time Comrade Captain manoeuvred his ship close to the rig and the KGB press corps fired off their shutters, the men were lined up along the rail peeing over the side and waving their hard hats. No matter what dodge the Russian tried and whatever angle the rig was photographed from, the men were waiting with everything hanging free. It was the finest and most effective display of international weaponry ever raised against the Soviets! With a final, exasperated circle round the rig the trawler headed away over the eastern horizon.

Perhaps news of the secret deterrent sent a cold shiver of apprehension through the Russian Navy. The following day a submarine surfaced near the rig, and a solitary figure on the conning-tower studied us intently through binoculars for a few minutes before the sinister-looking craft sank from sight below the surface.

According to hydrographic statistics, the height of the waves in the North Sea decreased in spring and summer, though when they crashed against the rig at the height of a gale it was difficult to believe. Periods of very strong wind were certainly less frequent and, as on land when after a long winter hibernating animals stir into life, now the North Sea was wide awake with the frantic activity of oil and gas exploration and development. Heavy lifting barges like the *Hercules* were towed out to locations far from land to build platforms over new production wells.

Large ships arrived in Yarmouth with thousands of tons of casing and drill pipe, which in turn were loaded onto the supply boats and shipped out to the rigs. Stockyards were piled high with massive pipes, encased in concrete, ready to be taken out to the giant pipe-laying barges preparing to inch their way from the Hewett and Arpet production wells to the new gas terminal at Bacton, on the Norfolk coast. Boats of every size and description were pressed into service to carry stores, men and equipment backwards and forwards, and the skies hummed with helicopters on the same mission.

Some of the small craft were neither designed nor equipped to go to sea and frequently got into difficulties. Two of Phillips Petroleum's top executives had an experience they are never likely to forget, when they arrived in Yarmouth on a whistle-stop tour of their European operations and asked to be taken out to the Hewett. Unfortunately it was a day when a typical east-coast fog had brought the heliport to a standstill but, undaunted, they insisted on going by boat, and rather than disrupt the schedule of the supply boats, a local man, who owned a fast ex-German motor-torpedo launch, was chartered to take them.

In normal conditions it should have reached the rig in under two hours, but the boat was not fitted with radar, and the skipper's knowledge of navigation was limited to taking visitors on short fishing trips off the harbour entrance. He steered through the fog but made no allowance for the strong tidal stream, and four hours later, hopelessly lost, he radioed the Hewett asking for help. There was little I could do other than ask our standby boat to locate him on radar, but there were so many trawlers in the area that it was impossible to pick out the launch among them. To add to the problem, the Phillips executives were tired, hungry, thirsty and very angry.

'I wanna speak to Bill MacMillan,' snapped one of them over the radio.

Bill was standing by my elbow and paled when I handed him the microphone. He pressed the transmit button.

'Yeah, Carl, I hear ya,' he said nervously.

The voice from the boat was almost hysterical: 'Bill, you gotta get us off this puddle-jumper, and I don't give a damn how you do it!'

Bill gulped. In the tone of the man's voice there was a veiled threat. What could he do? Beyond the radio-room visibility was a hundred yards at the most, and he dare not send the rig standby boat to search for the launch. The power of the Phillips executives was mighty, but the British Government's regulations were mightier, and the rig could not be left unprotected. Bill was in a quandary and had to play for time.

'OK, Carl,' he said confidently, 'we're working on it.'

Though the voice may have sounded confident, the expression on his face was one of helpless despair. Ricky made it clear that it was none of his concern and hurried out on deck, quickly followed by the driller who had been discussing the drilling programme with him. Bill lit a cigarette and blew a cloud of smoke at the ceiling.

'Jesus Christ!' he cried. 'What are we gonna do about those guys?'

He was actually speaking to me, but the great oilman in the sky obviously thought the appeal was directed at him, and in his compassion the response was almost instantaneous. Had he replied personally, he might have done so in the style of Scotch Jimmy: 'Rejoice with me, for I have found my sheep which was lost. Luke, Chapter 16,' but instead he spoke over the radio through the excited voice of the launch skipper.

'I can see the rig! I can see the rig! The fog's lifting. You're about three miles away.'

The two aged Phillips men were scooped off the launch in the personnel basket, and Bill MacMillan ushered them into the galley to smooth their ruffled feathers with large steaks, hot coffee and fat cigars. Apart from the senior one of the pair coming to the radio-room to get me to put a call through to his wife in Oklahoma, I saw little of them. They were obliged to stay overnight and left early in the morning on a specially chartered helicopter.

Trawler-owners operating out of Lowestoft and other ports along the east coast of England were angry about losing many of the traditional fishing banks as prospecting for gas and oil was intensified during the 1960s. To them it was just another nail in the coffin of the British fishing industry, hammered in by Westminster politicians, content to see fleets of trawlers go for scrap while allowing the Icelandic, Norwegian, Danish and Dutch boats to flood the UK markets with fish. Grumblings of resentment towards the rig-operators grew, with accusations that scrap metal, discarded wire rope and other rubbish were being dumped indiscriminately onto the sea-bed, causing loss and damage to trawl nets.

The explosive operations carried out by seismic survey vessels caused a tremendous furore, and I once sat in a Lowestoft pub with a group of trawlermen who were planning a mass protest. As the night wore on and gallons of strong ale poured down salty throats, the schemes went beyond letters to the newspaper to throwing cods' heads from the public gallery in Parliament, harassing the seismic ships at sea and, if all else failed, ramming one and sinking it. Fortunately in the cold light of morning, when the alcohol fumes had dispersed, letters to newspapers and Members of Parliament seemed the only sensible - though as it turned out, totally ineffective - recourse.

When all the countries bordering the North Sea first got together and divided the area into blocks, and licences were granted in the British sector 'to search and bore for and get petroleum', the oil companies needed to know with some certainty where oil and gas might be found in their particular block before embarking on the expensive business of drilling. Specialist geophysical exploration companies were employed and several different methods used.

One was to fly over the area with an aircraft fitted with an instrument which recorded the magnetic properties of the rock, even though it was under the sea. From a magnetic map produced by the instrument, a geophysicist could determine where the possible oil- or gas-bearing regions were. It was a very useful device, but an even more accurate method, and the one which enraged the fishermen, was the seismic survey. It required two boats. The first, usually a large vessel of several thousand tons, towed a buoyant cable, about 1½ miles long, attached to which were over a hundred sensitive pieces of equipment capable of picking up signals. As the recording boat steamed along a set course, it electrically exploded fifty pounds of dynamite (which had been laid by the second boat) under the water. Sound-waves from the explosion penetrated deep into the sea bed, then bounced back up to instruments on the recording ship by way of receivers suspended along the cable. It was an accurate and effective method of interpreting geological structures, but hard luck on any shoals of fish that happened to be in the proximity of fifty pounds of dynamite when it went off.

Seismic boats could cover up to sixty miles a day, firing off dozens of charges, and the fishermen protested that thousands of fish were being killed or maimed. The scientists countered by accusing the fishermen of being alarmist, and insisted that damage from the shock-waves was minimal. Who was right was never decided upon, but I do know that whenever seismic ships were operating in the area, even several miles away, they radioed the rig with a warning that divers should not go below the surface; and after the explosions scores of dead fish drifted past on the tide.

Day-to-day life on the rig had hopes and ambitions, troughs and highs, pleasant and not-so-pleasant times, like life anywhere. For a period the days would come and go without any mishap. Then, as if the gods had yawned with boredom and decided it was time for a bit of action or drama on the Hewett, things would start to go wrong. The drill bit would stick, an essential motor run itself to destruction, a vital piece of equipment fail to arrive, or when it did, a wire sling would snap as it was being lifted off the boat and it would fall into the sea.

At the beginning of May, during my first night on the rig after shore leave, I was shaken out of a deep sleep by a roustabout jabbering that there had been an accident in the engine-room. I pulled on boots, trousers and sweater, grabbed my first-aid bag and raced after him down the flights of stairs to the depths of the rig, where I found a man lying on the engine-room floor writhing in agony, with blood pouring from a horrible wound. His right thumb had been severed through the bone and was dangling from his hand by a thin strip of skin. Even above the noise from the big diesel engines I could hear the man screaming with pain.

Injecting him with morphine, I replaced the thumb in position as best I could with a splint and a dressing, then, with the help of the roustabout crew, carried him up to the radio-room. He was in a severe state of shock, but eventually the morphine took effect and put him to sleep. Fortunately the weather was good, and although it was 4 a.m. when I called for a helicopter, within two hours the man was on his way to hospital. By a miracle of surgery the thumb was sewn back on, and though it took a long time, the bones apparently knitted together again. But the man never came back to the rig.

It is often said that troubles come in threes, and they came in rapid succession that week!

Late the following evening I was in the radio-room talking to Larry and helping him to write to Rose-Ann. Her latest letter had

been full of news about how all the folks back home were preparing for the wedding. An old neighbour had died; his twenty-
acre farm was vacant, and the wealthy city-man who owned it
said Larry could rent it, provided he was home before the end of
the year. Rose-Ann's uncle had promised to give them a mule
for a wedding present, and other neighbours had offered help to
get them started on the farm. Larry could hardly contain himself
with excitement, and we worked on a long letter, telling her
about a trip he'd made to London when he was ashore. Suddenly
a voice blared from the aircraft frequency radio.

'Hewett "A", Hewett "A", this is a BEA helicopter. I have an
emergency! I must land on your rig.'

'Get Vernon quickly,' I called to Larry. 'Tell him to stop the
cranes.'

Larry shot out of the room, and I had hardly time to switch
the landing-lights on and give the helicopter clearance to land
than there was a tremendous thump on the ceiling above my
head that shook the whole rig. Vernon charged outside and a few
minutes later returned with two white-faced pilots and twenty
passengers. The helicopter had been on its way to a rig when the
oil feed to the main rotor blades had failed. With only five minutes flying-time before the gearbox seized up, providence had
placed the Hewett under its flight path.

In times of emergency the bitter rivalry between competing
oil companies was put aside, and although the huge helicopter
clogged up our helideck and disrupted crew-changes, Phillips
very generously laid on meals for the passengers and a boat to
take them to Yarmouth. The pilots stayed on the rig awaiting
help, and at daylight a small helicopter arrived and hovered
while two mechanics and spares were lowered onto the rig.

For a few hours the situation was very tense. If the main
gearbox was seized totally, it would create major problems. It
might take many days to replace it, and meantime the only access to the Hewett would be by sea. That was no problem provided the weather remained stable, but if a gale blew up and
there was an accident, casualty or other emergency, what then?
These thoughts must have rushed through the minds of both
Vernon and Bill MacMillan as they stood grim-faced watching

the mechanics unscrewing inspection hatches and pushing the rotors round by hand. To everyone's relief the mechanics announced that the damage was not as catastrophic as was first believed, and having fitted new parts to the oil system they sat with the pilot while he started the engines and ran the rotors at high speed for several minutes. Toolboxes were heaved on board, and with a wave of thanks from the crew the Sikorsky lifted into the air and set off for Yarmouth, flying a few feet above the sea.

The third mishap occurred shortly after the Sikorsky had left. A supply boat arrived from Yarmouth with a crew-change, and Ron Gower, the crane-operator, was busy lifting the men onto the rig with the personnel basket. About ten men had been landed and Ron was swinging the jib away from the rig for another load when, without warning, the wire rope parted from the crane and, together with the personnel basket, plunged into the sea and sank. It was a miracle that there were no men on the basket at the time. The boat had to move to the east side of the rig, where the second crane and the spare personnel basket lifted the rest of the crew aboard. When the faulty crane was examined, it was discovered that corrosion caused by salt spray and sea air had eaten through the bolts which fastened the rope to the crane's winch drum - something that would have been detected only during an annual inspection.

The merry month of May, when in villages throughout England people danced around maypoles to welcome the beginning of summer, was wet and uncomfortable on the North Sea rigs, and to make things worse there was an outbreak of 'crabs' on the Hewett.

Most of the men were fastidious about having a shower after each tour, but there was one revolting individual who rarely washed or shaved and wore the same underclothes and shirt for the two weeks he was on the rig. No amount of cajoling from his mates had the slightest effect on the imbecile, and encouraged by the hot, sweaty confines of the cabin, it was inevitable that an infestation of lice would break out in what the *Ship Captain's Medical Guide* described as 'the hairy part around the privates'. His closest friends reacted as if he had contacted the plague and swiftly vacated the cabin, but it was too late.

The lice spread like a forest fire, and I was besieged by men, hands in trousers, scratching feverishly, but there was nothing in the medical cupboard to halt the noxious beasties. In desperation some of the men sprayed the offending parts with diesel oil or rubbed in salt. Others tried engine-room grease or, if the chef was in a benevolent mood, margarine or corn oil. None succeeded in stopping the itching, and some of the more volatile remedies worsened the condition by causing painful rashes.

I radioed Yarmouth hospital for assistance and spoke to a nursing sister, who could hardly stop laughing but promised to send a supply of cream which she assured me would be effective. Two parcels arrived on the next helicopter. One contained several jars of antiseptic cream with typed instructions which read: 'Lift male organ with clean dinner fork and apply cream liberally to surrounding hair with a toothbrush - do not use for brushing teeth. Should the cream fail to cure the complaint, resort to treatment in parcel number 2.' I tore open the second parcel and found a spent twelve-bore shotgun cartridge with the message:

'Supplies of these are available freely from any gunsmith's. Fire no.6 shot at Texan crabs, but any smaller shot adequate for British.' Luckily the cream helped to check the lice, and drums of DDT powder issued later to every cabin for shaking into bed-clothes finally killed them off. But Ricky was furious.

'Which sonofabitch brought that theyar mechanized dandruff on the rig?' he bellowed, when the roustabout crew were assembled on deck. They shuffled their feet and stared into space but no one spoke. Ricky glared at them. 'OK, you bunch of chicken-shit. I'll put it to ya this way. You guys are scheduled to hit the beach on tomorrow's crew-change, and if you ain't talked 'ole itchy ass into finding another job by the time you're due back, the whole goddam bunch of you are fired!'

He stormed away and the roustabouts went into a huddle to discuss what should be done. I never did discover how it was resolved, but when the crew arrived back a week later, ole itchy ass was missing.

The main preoccupation of the men cooped up on the rig for two weeks was women. Cabin walls were plastered with pages torn from girlie magazines sold by the crews of the German and Dutch boats, and pornographic films and photographs were traded like cigarette cards. Jokes were invariably about women or sexual frustration, and a favourite one making the rounds concerned two rig workers talking together.

One said to the other, 'What's the first thing you do when you're off the rig?'

'I get a woman into bed and get my leg over,' he replied.

'Well, what's the second thing you do?' asked his mate.

'I put my suitcase down and take my coat off.'

The regular invasion of Yarmouth by a large number of randy men with plenty of money in their pockets quickly attracted ladies of easy virtue. Sometimes they were no more than school-girls escaping from an oppressive home life. With their few possessions crammed into tatty suitcases, they hitch-hiked across the country in search of the bright lights and a good time, working in hotels by day and living it up at night.

The experienced prostitutes were very different, usually well dressed, reeking of cheap perfume and plastered in make-up. Very often they worked for a scarfaced pimp who pocketed the takings and roughed up any client who would not pay or became violent with one of the girls. They operated from outwardly respectable hotels and could be hired for private parties or to take part in bizarre orgies. However abhorrent the presence of these girls may have been to the moralizers, there was no doubt that, had they not been available, no local girl would have been safe.

Other countries, less 'civilized' than our own, recognize the important role of regulated prostitution and protect the health of the country by ensuring that the girls are registered and have regular medical inspections. In Britain the head-in-the-sand do-gooders who have influenced our Governments - often the same highly placed people whose names appear in newspapers when a sex scandal is uncovered - choose to evade the issue.

As a result of this short-sighted policy there was an outbreak of venereal disease on the rig. Men who regularly indulged in the seamier side of Yarmouth's nightlife were in danger of contacting it, and very often did. On the rig there were many men who carried the disease and either concealed it or sought medical help privately. It was a miracle that there were only very few reported cases among rig workers of that even more virulent strain of venereal disease, syphilis. Had that taken hold in the confined living-quarters of a rig, it would have spread rapidly through guilty and innocent alike. Venereal disease would have been unheard-of on the rigs had not British Parliamentarians, aided and abetted by the Church, adopted a prudish Victorian 'We are aware it goes on but we decline to discuss it' attitude towards the nation's sexual habits.

June was a superb month, with a cloudless blue sky day after day and not the slightest breath of wind. Off duty the men sunbathed on the helideck or fished over the side of the rig with long lines baited with bacon fat. The *Ross Delight,* our standby boat, tied up alongside, and I showed the skipper and crew round the rig, then joined them on board the boat for a can of beer and a yarn. They were glad of a change from the boredom of lying at anchor or steaming up and down until at the end of three weeks the *Smit Lloyd 6* or the *Inkenturm* took over standby duty while the *Ross Delight* sailed away to Yarmouth to change crews.

Even in good weather it was a tedious existence for men used to the excitement of trawling, and in bad weather a hell of a life. When there was a gale-warning and every other boat was running for shelter, they had to ride it out and face what nasty weather nature hurled at them. If, as so often happened, the end of a tour of duty coincided with the onset of a fierce gale, and the relief boat could not leave harbour, the trawler crew could be at sea for a month or more. Theirs was a thankless task that lacked sufficient kudos to attract the attention of the media, yet to the men on the rig the sight of the sturdy little boat lying nearby, or the gleam of its lights at night, was very reassuring.

Phillips took advantage of the good weather to stock the rig with fresh supplies of drill pipe, casing and essential equipment, and there was a flurry of activity as supply boats arrived and departed at all times of the day and night. A British boat was chartered to assist the Dutch *Smit Lloyd 6* and the German *Inkenturm,* and it was interesting - and, for anyone clinging to the belief that Britain was still a great nation, painfully embarrassing - to compare them with one another. The Dutch boat was so immaculate that it always looked as if it was brand new and had just been launched from a shipyard. Below deck, even in the engine-room, everywhere was clinically clean and neat, and in the wheelhouse the instruments and controls were polished to the standard of a luxury motor yacht. The crew always wore clean coveralls and took an immense pride in looking after their ship. Teutonic pride and discipline were equally evident on the German ship, with not a speck of rust, patch of oil or unkempt crew member to be seen. But when dear old 'Rule Britannia' arrived on the scene, what a contrast!

British boats epitomized Masefield's romantic image of a 'dirty British coaster with a salt-caked smoke stack, butting through the Channel in the mad March days': they were always dirty and salt-caked, whether it was mad March or flaming June.

When a British boat arrived at the rig from Yarmouth, an innocent observer might easily have been deceived into believing that it had just sailed non-stop around the world. Hull, wheelhouse and deck were streaked with rust, and a ragged, smoke-grimed ensign that had once been red hung limply from its mast. Down below, the smell of greasy fried bacon pervaded the accommodation, and empty beer cans rolled around the crew-room floor. Why, as a race, the British are slovenly would probably need a psychologist to explain, but perhaps, having lost the Empire, personal and national pride have gone as well.

The sea has always been looked on as the biggest dustbin in the world, and every nation is guilty of polluting it. It is odd that the people who earn their living on the sea are the worst offenders. Tanker captains blatantly clean out their tanks in mid-ocean; the crews of work boats and cargo vessels of every size dump surplus scrap, empty paint cans, glass bottles, oil drums, plastic sacks and containers without a moment's hesitation. Fishermen often accused the rigs of indiscriminately dumping scrap metal and wire rope which damaged nets and trawl gear; but, while there is no denying that it happened, it was due to some irresponsible workers rather than company policy.

Both toolpushers on the Hewett were very strict about what they referred to as 'good housekeeping', and garbage that was not bio-degradable was placed in skips and loaded onto a supply boat to be taken to Yarmouth. But, despite the vigilance of the toolpushers, there were a few roustabouts too idle to put the lid on a can of paint or clean a brush when they finished work, and they would calmly drop them over the side.

One trio of surly, uncouth thugs who worked on Ron Gower's roustabout crew were continual trouble-makers and put more effort into avoiding work than actually doing any. When told to load a skip with junk, they would heave everything over the side, including the skip, if no one was looking. If they were given the job of painting the guard-rails round the edge of the rig, as soon as the crane-operator's back was turned they would dab a few smears of paint on the metalwork and at the same time ease the can of paint over the side with their feet. They thought it a huge joke and were able to waste an hour going to the store for more paint. They made themselves unpopular wherever they went, and everyone hoped they would quit or be fired, but they always managed to survive.

When the end came for them, the hero of the day was Barry the welder. They had been sent to help him move his welding equipment and some sheets of steel down to the cellar deck below the drill floor, and as usual they were reluctant to work. Impatient to get the job done, Barry lugged most of the gear down to the cellar deck himself and was about to start welding when the trio decided it was time to slink off for a smoke.

'Hey, you can't go yet!' Barry exclaimed. 'Another five minutes and we'll have the job done.'

'Stuff you and your job!' snarled the bull-necked gorilla who always assumed leadership. 'We're going for a smoke. Get out of the way or I'll move yer!'

He clenched his fist and threw a punch at Barry's head, and that was the biggest mistake of his life. Barry had been a fairground boxer turned professional wrestler before working in the shipyards and was as hard as nails. Neatly side-stepping the punch, he drove his hairy fist into the man's gut and followed it up with another crashing blow to the jaw that laid him out cold. With howls of rage, the other two snatched up steel bars and rushed at Barry with murder in their eyes, but he stood his ground. Before they realized what had happened, Barry had both of them clamped by the neck under his brawny arms, and he squeezed harder and harder until their faces turned purple and they gasped for mercy. The vice-like arms released their grip, and the two men dropped like rag dolls to the floor grating. Picking up his welding torch, Barry calmly finished the job as if nothing had happened.

Happily for all, Bill MacMillan had witnessed everything, and the trio were ordered off the rig to a supply boat that had actually left for Yarmouth but was recalled in order to get rid of them. Rimrock's personnel officer moved fast, and when the crew-change helicopter arrived the following day, there were five new roustabouts on board. Graham, a British lad in his early twenties, two Frenchmen, a Greek and a Turk. They were hardly given time to change into coveralls before they were scraping rust and wielding paintbrushes on deck.

After a few days painting on deck, the Greek was lowered down into a deep, narrow tank to clean it out and somehow managed to get his foot wedged between two pipes. His mates heard his cries for help and lowered him a rope, but though they pulled and pulled, the poor Greek was stuck fast and screamed with pain. Vernon was quite unsympathetic.

'Quit hollerin' and get your ass out of that theyar tank!' he shouted, as if expecting him to pop up on command, but the man only moaned with fear and prayed aloud in Greek.

The afternoon went by and, though everything bar cutting his leg off had been tried to release him, as darkness began to fall the unfortunate Greek was still in the tank. It had been a hard day for Vernon, dealing with a series of urgent problems, and he was lounging back in his chair with his heels resting on the edge of the desk, rolling a cigarette, when Ron Gower, the crane-operator, rushed into the radio-room panting for breath and spattered with water.

'Vernon,' he gasped, 'what are we going to do about this bloody Greek?' The poor sod's been there all day. How are we going to get him out?'

Vernon sighed deeply and pushed his hat to the back of his head.

'Aw, shit,' he drawled irritably, 'put the goddam Turk in with him!'

Whether this explosive combination would have worked, we never did find out. The Greek's wailing incantations to his god had been answered, and he had dispatched one of his emissaries under cover of darkness to instruct his countryman to untie his boot-lace and remove his feet from his boots. Lo and behold! he emerged from the tank in time for breakfast.

In the time I had worked on the rig I had learned to size up people very quickly, and there was something about the two young Frenchmen that was distinctly strange. It was not that they were unfriendly or morose, it was the habit they had of asking furtively if anyone had been enquiring about them, and appearing at my cabin early in the morning or late at night for lint and sticking-plaster to dress a wound. One of them explained that he had been in a car crash and that the hospital had told him to change the dressing every day. He would never allow me to see the wound, insisting that it was only a scratch and had almost healed. A week later, I was shaken awake in the middle of the night to find one of the Frenchmen slumped on the spare bunk and the other standing over me.

'Keep quiet, monsieur,' he whispered. 'We wish to speak to you. Can we trust you?'

I yawned and rubbed the sleep from my eyes.

'Can't it wait till morning?' I mumbled. 'There's some first-aid stuff in a box outside if that's what you want.'

'No, no, monsieur! We must talk. Can we trust you?'

It was obvious that I was not going to get any sleep, so I sat up.

'It depends what you want to tell me,' I said. 'What's your problem?'

'You must help Raoul,' he whispered. 'He hurt bad.'

'Well, why didn't he stay in hospital then?' I said irritably, wishing they would go away and let me sleep.

'No, no, Monsieur, you don't understand.' He turned to Raoul and spoke rapidly in French. Raoul looked at me, then back at his friend. '*Oui*,' he said sullenly. 'We are from Corsica and run from ze police,' the Frenchman announced bluntly.

I groaned inwardly. Life on the rig was becoming too complicated. What had I done wrong that fate should pick on me to send two Corsican bandits half way across Europe to wake me up in the middle of the night on a rig in the North Sea?

'Well, if you've killed someone,' I said wearily, 'there's no way I can help you.

'No, no, no, we no kill! Ze police round up all ze Corsicans in Paris, so friends get us passports and car and we go to Belgium, but car crash and Raoul hurt bad. You look!'

He spoke in French, and Raoul rose stiffly to his feet and opened his shirt. On his right side, below his armpit, a large lacerated wound had been clumsily covered with lint, and it was almost welded into his skin with dried blood and oozing pus.

The area around it was bright red, and when I touched it, he winced with pain.

'My God, what a mess!' I exclaimed. 'How long has it been like this?'

'Maybe one, maybe two days,' whispered Raoul. 'You fix it, hey?'

'No chance, mate,' I said. 'You need a doctor quickly or you'll die of blood-poisoning.'

At the mention of a doctor the other man stared at me with menacing dark eyes.

'Monsieur,' he said flatly, 'you will not tell ze doctor or any- one, OK? You please fix Raoul and do it good!'

I felt that if I refused I might get a knife in my ribs, and as they were due to go ashore on the crew-change helicopter within the next day or two, there was a good chance they would not return. I swung out of my bunk and pulled on trousers and a sweater.

'Take your shirt off,' I said to Raoul. 'I'll be back in a few minutes.'

As I opened the cabin door, the other man laid his hand on my shoulder. 'Remember, monsieur, no doctors!'

Returning with my first-aid bag and a bowl of warm disin- fected water, I spread a rubber sheet on the spare bunk and got Raoul to lie on it. After repeatedly soaking the congealed lint I was able to free it, but Raoul gasped with pain and gripped the side of the bunk when I slowly peeled it off. Swabbing away the matter, I searched for signs of metal lodged in the wound, but it appeared to be clean and, dusting it with antibiotic powder, I covered it with a dressing.

After the Corsicans had gone, I went to the galley for a mug of strong coffee, then lay on my bunk wondering whether I should just forget about them or risk a knife or a bullet to let the police know, but I fell asleep without reaching a decision. To my surprise, on the morning of the crew-change they bounced into the radio-room as lively as ever, and Raoul pulled his shirt up to show me the wound. Amazingly, the antibiotic powder had worked fast and the inflamed area was almost gone. I changed the dressing, and as they were about to go outside to the helicopter, Raoul grabbed my hand and shook it vigorously.

'You a good guy!' he beamed, handing me a scrap of paper with an address on it. 'If you ever in trouble and a guy say he kill you, let me know. Wherever you are, I come, and' - he formed his right hand into the shape of a gun -'bang, bang, he dead! OK?'

'OK,' I laughed.

Little did I realize that he was serious. A short time later I was idly browsing through a newspaper in a Yarmouth barber's shop while I waited to have my hair cut, when the headline 'French Police Gun Down Bank Raiders' caught my eye, and I read: 'French police killed three bank raiders in a gun battle on the outskirts of Paris. The leader of the gang was a twenty-four-year-old Corsican, Raoul Dubois, who was wanted for questioning about a number of serious offences.' All the antibiotic powder in the world would not have saved the handsome man with the flashing smile on that occasion.

When I returned to the rig after my two weeks off in July, the news was gloomy. The rig mechanic's son had been killed in Vietnam, and his son-in-law was missing after a helicopter had crashed in the jungle. It deeply affected the other Americans, who also had relatives out there, and instead of the normal banter and jokes at mealtimes they sat in silence, chewing languidly on their food. There was an air of despondency and hopelessness about the rig, as if the men had lost their zest and had left the drill to revolve aimlessly on its own.

Charlie the roustabout had been discovered lying on deck within inches of falling into the sea and had been sent home to recover from an attack of high blood pressure and giddiness; he could tell a convincing lie, Charlie could!

The only spark of light in the gloom was the irrepressible Larry, who appeared at the radio-room door grinning happily and clutching the latest letter from Rose-Ann. As ever, she was bubbling over with excitement about the approaching wedding. She said she really would have liked to have been married in summer, but as Larry couldn't get home until after the fall, her Ma had asked the preacher to fix it for Thanksgiving Day in November, so it would have a double meaning. Larry's folks had been asked if it would be OK with them, and Larry's Pa had said, 'It didn't matter nohow what day it was. Gettin' married and livin' in a shack was like dyin'. Once a woman got a man inside the wooden box, there's no way he'd ever git out agin!'

Her Uncle Jeb, who at seventy-five had married May-Lou, who was eighteen, had died after only six months of marriage. 'When I asked old Doctor Benson what Uncle Jeb had died of,' wrote Rose-Ann, 'he said it was too much pokin' and not enough breathin', and I sure don't know what he means by that!'

Larry giggled like a schoolboy when I finished reading out the letter. 'Gee, I guess for all her schoolin' there's a coupla things I'm gonna have to teach Rose-Ann,' he sniggered. 'With

bein' an only chil', she had her own room and I guess she didn't
see nothin'. Ma and Pa used to keep us kids awake nights when
they was humpin', an we use to watch 'em to figure out what the
hell was goin' on.'

The pervading uneasiness on the rig made everyone very
touchy, and Ricky was in a permanent bad mood, firing men in
the morning, then in a pang of remorse re-instating them in the
afternoon. He rampaged round the rig like a bear with a sore
head and upset little Yorkie, the assistant mechanic, by calling
for all engines to be run to get maximum power on the drill
floor. Yorkie tried to explain that number 6 engine provided
electricity only for the domestic supply on the rig, but Ricky
wouldn't listen.

'Don't give me no shit!' he yelled. 'Ah want every sonofabitch rollin'.'

Yorkie threw his hard hat to the floor with exasperation. 'If
you can hang on a minute or two,' he yelled back, 'you can have
the fucking washing machine running as well.'

Out on deck, the roustabouts were busy unloading a supply
boat, and although Ron Gower was one of the most experienced
crane-operators in the North Sea, Ricky decided he was doing it
all wrong.

'Ya too goddam slow!' he bawled. 'Bust that load of pipe off
the boat and get it up heah. We ain't got all day to waste.'

'You can't snatch a load as heavy as that,' protested Ron.
'There's twenty tons in each lift; it has to be lifted slowly.'

'Don't tell me what ah gotta do and what ah ain't gotta do!'
thundered Ricky. 'Bust it out when ah tell ya!'

'Not bloody likely!' said Ron. 'It could pull the crane off the
side of the rig, and me with it.'

Ricky's face went purple with temper. 'Goldarn it, ya either
do as ah tell ya or ya fired!'

Ron shrugged his shoulders. 'Suit yourself. There's always a
demand for good crane-operators. I won't be out of work for
long.' He climbed down from the cab of the crane and walked to
the accommodation block, pulling his coveralls off as he went.

'Goddam smart-ass Limey!' snorted Ricky at anyone who cared to listen. 'I'll show the sonofabitch how to ride a crane.' Heaving himself into the cab, he started the engine and, slewing the jib out over the boat, dropped the weighted hook like a missile. 'Flip that sling on,' he roared at the seamen.

One of the crew slid a wire rope sling round a stack of steel pipes, dropped it over the crane hook and was almost knocked overboard as Ricky revved the crane engine at full speed and snatched the load into the air. The crane shuddered and swayed alarmingly as the heavy load shot up to the rig platform and was slung inboard. In his haste, Ricky misjudged his timing and slewed the jib so fast that twenty tons of pipe crashed into the top of the accommodation block, demolishing a large section of the steel safety-net around the helideck and flattening my radio antennas. Ricky sat in the crane cab and stared, crestfallen, at the carnage.

'Holy shit!' he groaned. 'What happened?'

The roustabout crew, sensing that heads were about to roll, made themselves scarce, but it was a very subdued Ricky who stepped down from the crane and walked slowly across the deck to the mess-room to find Ron Gower. All the braggardly boasting had gone.

'Say, Ron,' he said, slipping self-consciously into a chair, 'Ah guess we've bin' buddies for a while now, and ah've bin thinkin' . . .' He stopped to see if there was any reaction, but Ron ignored him and continued to drink his coffee. Ricky cleared his throat. 'Yeah, ah've bin' thinkin', y'ain't such a damn bad crane-operator after all.' He slapped Ron's back, and his face glowed as he thought of more praise. 'Why, ya nearly as good as an American!' He realized what he had said. 'Hell no, ah mean there's some American crane-operators who ain't as bad as you.' He had got it wrong again and scratched his head in annoyance. 'Oh, Jesus, what am ah sayin'?' he mumbled. Ricky was never very articulate when he was trying to be kind. Suddenly his face hardened into the familiar scowl. 'Ron,' he growled, 'get your ass into that theyar crane and offload the boat!'

There was the merest flicker of a smile on Ron's face as, without a word, he drained his coffee and went out. Before it went dark, the deck was neatly stacked with pipe, the welder and roustabouts had repaired the helideck and with the help of Larry I had fitted new antennas. The rig buzzed with laughter at Ricky's accident, and it helped to drive away the gloomy atmosphere. Even in the face of personal problems and tragedies, life had to go on, and there was no one who understood that more than men who worked in the offshore drilling industry.

Graham, the British roustabout who arrived on the rig at the same time as the Corsicans, was very keen to become a crane-operator, and whenever he had time Ron Gower taught him how to operate the controls. But what interested me about Graham was that he had served an apprenticeship in one of the top boat-yards in the south of England, building magnificent wooden yachts. He had masses of photographs of yachts he had worked on, and whenever he was off duty in the evening we spread a boat plan on the radio-room floor and worked out how much it would cost to buy the materials and build it. Scotch Jimmy was very scathing about 'wee toy boats' but liked to join in and describe the great liners he had watched being built on the banks of the Clyde.

Of all the interesting craft to be seen around the rig during July, the most fascinating was the gigantic pipe-laying barge, the *Hugh W. Gordon*, chartered by Phillips from Brown & Root to lay a gas pipeline from the Hewett to the terminal at Bacton, the nearest point on the Norfolk coast. Four hundred feet long by a hundred feet wide and with a crew of over 200, it was the world's largest pipe-laying barge and a floating continuous production factory, in which the forty-foot-long, sixteen-inch pipes, that had been coated with bitumen and concrete in Yarmouth, were welded together and lowered into the sea. Every eight minutes throughout the day and night, a hooter sounded on deck signalling that another length of pipe had been welded on and that the barge was about to move forward another forty feet, following a line of buoys laid down by a survey boat. The barge pulled itself along the line by anchors lifted and dropped by two American tugs in constant attendance, and the radio airways were alive with strange instructions from the controller on the barge.

'OK, buster, pick up the jewel and throw it some more' was an instruction to a tug skipper to lift an anchor and move it along a few yards.

I often watched the operation from the grating outside the radio-room, and Pat Malloy, an Irish roustabout, joined me one afternoon and we stared in wonder as the barge inched its way over the sea, discharging the pipe from the stern like a giant sausage-machine.

'It's marvellous, isn't it!' said Pat quietly. 'You've got to hand it to these Yanks. They might be full of bullshit at times, but for anything to do with drilling and getting gas or oil to the shore, there's nobody to beat them!'

'You're absolutely right, Pat,' I said. 'There's no country in the world that's got the technology to match the *Hugh Gordon*. You get the feeling that it wouldn't matter how deep the sea was, if there was gas or oil there they'd find a way to lay a pipe to the shore.'

To mollify the fishermen and to prevent possible damage to the pipe if ships anchored near it, the oil companies had undertaken to bury the pipe a minimum of six feet, and more American offshore ingenuity was enlisted. Following behind the *Hugh Gordon* was another barge equipped with very high pressure water hoses which blasted a trench in the sea-bed, then, as the pipe sank into it, covered it over again.

At one stage, the *Hugh Gordon* was working within a few feet of the rig and there was a lively exchange of newspapers, girlie magazines and porn movies between the crews. The barge also carried a large stock of duty-free cigarettes and cigars which were made available, and though the grim-faced battalion at the Customs House in Yarmouth would have taken a dim view of the transactions, the men deserved some compensation for enduring the harsh conditions on the rig. The few pounds lost to the Treasury would not have kept a cosseted Government minister in luncheon-vouchers. However, the pleasure was short-lived. Either it was the extra-sensory disapproval of the Customs and Excise or simply an atmospheric phenomena that caused a sudden and dramatic change in the weather, but for whatever the reason the barometer plummeted rapidly, and within an hour of the BBC shipping forecast's predicting nasty weather in the North Sea we were being lashed by a gale, and the *Hugh W. Gordon* was hauled away into open water.

The gale was surprisingly vicious for July, gusting to force 9 (60 mph), and from our elevated view on the rig it was intriguing to watch enormous waves break over the barge and sweep across the deck. It seemed as if the heavy rolling and pitching would twist the pipe hanging over the stern and break it, but the clever engineers had thought that problem through, and in heavy weather the end of the pipe was suspended alongside the barge from davits.

During the night the screaming wind and booming of the waves against the rig kept me awake, and I was reading in my bunk when there was an urgent hammering on my cabin door and a figure swathed in oilskins, dripping with water, rushed in. It was Graham, the roustabout, and he was shaking like a leaf.

'Bob, come quick!' he panted. 'There's been an accident. I think the guy's dead!'

I leapt into my clothes and, grabbing my first-aid bag and the folding stretcher, ran with Graham to the rain-soaked deck, where the roustabout crew were grouped round a body lying beside a stack of pipes.

'What happened?' I shouted above the screeching of the wind.

'He fell off the pipe stack,' bellowed one of the roustabouts.

Sliding the stretcher carefully underneath the man, the roustabouts carried him into the radio-room, and helped by Graham I cut his oilskin suit off. He was alive but there was a disturbing amount of blood oozing from the area of his belly, and when I pulled his shirt and trousers open, Graham had to rush out of the room to be sick. A large piece of intestine was hanging out of a long, jagged gash near his navel, and blood was flowing from the wound like water. Pulling on a pair of sterilized surgeon's gloves, I gingerly pushed the intestine back into the hole and placed a dressing over it to stem the flow of blood. Graham returned with Ricky, and while they looked after the man, I put an urgent call through to Yarmouth Hospital.

'Give him an injection of morphine before he regains consciousness and evacuate him immediately,' barked the doctor. 'Move the intestine if you can but don't do any more than let it rest comfortably in the wound, and cover everything lightly with a sterile dressing. Have you got a saline drip?' I said I had. 'OK, well insert it right away.' Whilst I was talking to the doctor, Ricky radioed the Phillips base and got the night operator to call Bristows to send a helicopter.

'Tell that chopper outfit to get a whirlybird out heah goddam quick. We gotta man with his guts hanging out!' he snapped.

The injured man groaned and started to regain consciousness, and I quickly injected an ampoule of morphine into his arm. Apart from the instructions I had received at my local hospital at home, I had never administered a saline drip but, gritting my teeth, I pressed the needle into a vein in his arm and held it in place with strips of sticking-plaster, suspending the bottle above the man. Then I covered him with a blanket.

Bristows organized a helicopter in double-quick time, and despite the appalling forecast the pilot took off into a strong wind and heavy rain. Within minutes his calm voice came over the radio.

'Hello, Hewett "A", this is Golf Bravo Delta Foxtrot. My ETA is 0350 hours. Can you switch your beacon on, please? - What's your wind speed at the rig?'

The wind-speed indicator on the wall showed gale force and even touched severe gale at times. I pressed the transmit button.

'Hello, Golf Bravo Delta Foxtrot, this is Hewett "A". The beacon's on, and wind is south-westerly thirty-five knots with occasional gusts to forty-two.'

'Roger. Many thanks,' said the pilot.

If designers of ships and rigs were obliged to live aboard their brainchild for a few weeks, they would realize that they had given little thought to the possibility that one day an injured person might have to be carried along passageways and through doors on a stretcher. Though the helideck was on top of the radio-room and reached by two short flights of metal steps, actually getting the stretcher up to the helicopter in the teeth of the gale without dropping it into the sea or aggravating the man's injuries took a full fifteen minutes and the combined efforts of the roustabout crew. Aboard the helicopter a crewman took charge of the saline-drip bottle, and the injured man was soon on his way to hospital. I went to the galley for a mug of coffee and sank exhausted onto my bunk. Evacuating the trawlerman in March had been difficult enough, but moving a man with serious internal injuries was a nightmare. The accidents I was expected to attend to were becoming more than I could cope with, and it was clear that properly qualified medics were essential on off-shore operations, no matter how much it dug into an oil company's profits. Fortunately the man recovered from his injuries; he took a less hazardous job ashore.

After a gale, the helideck of the Hewett was often littered with birds of all shapes and sizes, struggling in the last throes of life or just resting until they had regained sufficient strength to take to the air again. Most of them were starlings, sparrows and finches, with a sprinkling of gulls and in winter perhaps an ex-hausted greylag goose, but occasionally there was a colourful stranger which no amount of leafing through a bird-identification book could name.

I once caused a major ruffling of feathers among Norfolk or-nithologists by reporting the description of a bird which some thought was a rare red-flanked bluetail, a native of Finland, but it stayed only a short time before heading north to tantalize bird-watchers in Yorkshire. Very often a racing pigeon would drop out of the sky, stay for a day, then set off again.

At the beginning of August the rig was adopted by a bedrag-gled little thing which set up home on a girder under the heli-deck. The men christened it Harold, after Harold Wilson, the Prime Minister of the day, and though instructions were given that it was not to be fed, pieces of cake and bits of breakfast toast somehow found their way to Harold's roosting-place, and he became so tame that he fluttered about the rig like a parrot. For a change of scenery and diet, he often flew out to the standby boat or balanced on the mast top of a supply boat while it was being offloaded.

His was an untroubled and contented life, and he might have continued to enjoy it for many months had he not objected to a larger bird's landing on the deck above his perch. With a squawk of anger he swooped to attack but, sadly for Harold, the larger bird was a Whirlwind helicopter and with a 'whoosh' he was sucked into the air intake of the engine. Not designed to run on minced pigeon, the helicopter's engine spluttered and cut out, and the rotors slowly wound to a halt. As soon as it had landed, the crew going ashore leapt in with the eagerness that only those who have worked on a rig for two weeks can appreciate, and there were howls of disappointment when the pilot broke the news that the turbine blades had been shattered and the engine could not be started. It took the Bristow mechanic three days to repair the damage, and it completely disrupted crew-change schedules and delayed the arrival of urgently needed equipment. When the helicopter finally flew away from the rig, the down-draught lifted Harold into the air on his last journey, and a little heap of charred feathers fluttered down into the sea.

Charlie the roustabout was among the crew who had arrived back on the helicopter, and he looked very ill. His eyes had sunk into deep sockets, and he had lost so much weight that I hardly recognized him. Walking or talking was an effort, and drugs were having little effect on the excruciating pain in his stomach. It was obvious that he would have to give up his job, and in the seclusion of my cabin he broke down and wept uncontrollably. His wife's illness was worse; she had to be helped in and out of her wheelchair and needed constant attention. The neighbours had been very kind but they could no longer spare the time to look after her. His wife was terrified that if he asked his doctor for help she might be sent to a nursing home because he was too ill to look after her himself. I listened in silence and have never felt so powerless and inadequate in my life.

Charlie knew he was dying and that the end was not far away, and yet his one concern was for his wife. Weak and ill though he was, he stuck it out on the rig for his two-week tour, and as we shook hands on the morning he went ashore, I knew I would never see him again. A week later, too proud to be reduced to living on Social Security, and desperately worried, Charlie closed the windows, blocked the doors of his house and turned on all the deadly coal gas taps, and he and his wife died in each other's arms.

Under Ron Gower's guidance, Graham had become very proficient at handling the crane, and to his delight, when he applied for a job as a crane-operator on a rig drilling near the Dutch coast, he was accepted and departed on the next helicopter. Whenever he was off duty in the evening, the rig radio-operator allowed him to call me, and we chatted over the radio for a few minutes and talked about a boat he had started to build in his garden. The popular magazine *Yachting Monthly* had sponsored the plans for a very attractive little twenty-foot dropped-keel yacht, and Graham planned to build it from mahogany and oak and varnish the hull.

Much to Ron Gower's concern, Graham's place on the roustabout crew was taken by Francisco, a young Spaniard. There was no man on the rig more willing and cheerful than Francisco, but he was a walking disaster. Everything he touched seemed to fall down, blow up or break into a thousand pieces; when he walked round the rig, the mechanics hurried to slam the engine-room door, mud engineers shooed him away from their pumps, and drillers would not let him venture anywhere near the drill floor. Entrusted with a wrench that was the only one of its kind on the rig and sent to tighten bolts on a piece of equipment, he would be back within minutes wailing, 'Señor, I theenk the wrench he fall in the water.' Or if, as often happened when working up a ladder, he showered paint all over men working below him: 'I am sorry, señores. I turn can to see if much paint is left in bottom.'

A supply boat arrived with a deck cargo of large pallets stacked with sacks of chemicals, and Francisco was put aboard to lift the slings onto the crane hook. With the rolling motion of the boat, one of the sacks on the first pallet to be lifted had moved, and Ron shouted to Francisco to hold on to it. There was no radio contact between the crane and the boat, and at a signal from one of the seamen Ron lifted the pallet rapidly into the air.

Looking down, he noticed that Francisco had disappeared, and he cursed him under his breath for sloping off to have a quick beer with the crew. Angrily he slewed the jib of the crane over the rig, lowered the pallet to the deck and nearly fell out of his cab with shock. Francisco was standing on the edge of the pallet, still holding on to the sack as instructed.

When the rig mechanic asked Ron to change the starter batteries on one of the big diesel generators, he took Francisco with him. The batteries were heavy and in an awkward position, close to a bank of live terminals, and as the operation had to be carried out while the engine was running, Ron gave Francisco precise instructions.

'You hold that side of the battery and I'll hold the other, and then we'll slide it over, but for Christ's sake, keep your head well away from the generator. OK?'

'*Si, Señor,*' said Francisco eagerly. 'You pull, I pull, and battery he move. OK!'

Nowadays the hard safety hats worn on rigs are made from fibreglass but in the 1960s they were made from aluminium, and roustabouts had to be very careful when working near electrical installations. As they stooped to grasp the battery, Francisco slipped, and his head went between the live terminals. Instantly there was a tremendous flash of sparks, and Francisco was thrown across the engine-room. Shaking with fright and with wisps of smoke rising from his singed hair, he clutched his melted hard hat in trembling hands. '*S-s-señor* Ron,' he stammered, 'the h-hard hat, I th-theenk he no good any more!' Luckily for Francisco, the hatband inside the helmet was made of thick rubber.

The Spaniards were a hard-working, happy-go-lucky bunch, but apart from Carlos, who acted as their spokesman, none of them had a very good command of English, and it frequently led to misunderstandings and grievances. Julio was convinced that I was the person responsible for paying out wages, and pestered me for days about money due to him. He always came at a busy time, when a helicopter was due or a boat was being unloaded, and became more and more abusive until one day I finally lost my patience and bundled him out of the radio-room.

Later that day I happened to go down to the men's changing-room looking for Vernon, but the room was empty and I was about to go out when Julio burst in, yelling about his money.

'How many more times do I have to tell you?' I said angrily. 'Go to the Rimrock office in Yarmouth when you're ashore. I don't handle wages!'

Babbling a torrent of Spanish, he reached inside his coveralls, pulled out a long knife and advanced towards me.

'I want my money, or I kill you!' he screeched.

I backed up against a row of metal lockers, hoping I could find something to defend myself with, but the doors were shut tight.

'Put the knife away, Julio, or you'll end up in gaol,' I said, trying to sound calm, but he was not listening. His eyes blazed with hatred, and with body crouched he moved nearer.

'I want my money, or I kill you!' he repeated.

With my back pressed against the lockers, I steeled myself for his lunge, hoping I could kick him in the crotch before he got me with the knife. Beads of perspiration ran down my forehead and into my eyes, and I longed to wipe them away; but I knew that if I moved he would attack. Gripping the knife in his right hand, he was within a few feet of me when the door opened and Carlos stood on the threshold. In a second he sprang across the room and, dragging the knife out of Julio's hand, shouted in Spanish and slapped him repeatedly across the face. To my astonishment, Julio collapsed on the floor, blubbering like a child. Carlos reversed the knife and offered it to me, handle first.

'I apologize for the insult, *señor*. Please take Julio's knife and throw it into the sea. Julio is very worried. His wife just had new baby and his, how you say, landlord, demand many pesetas for rent and say he will put wife out of apartment if he does not have the money soon.'

Wiping the perspiration off my face with my shirtsleeve, I said, 'Thanks, Carlos, but I don't want the knife. Give it back to Julio. If you will come to the radio-room and explain what the difficulty is over his wages, I'll see what can be done about it.'

It transpired that Julio had not been paid for the overtime he had worked, and though Rimrock disliked parting with money at the best of times, a radio call to the company accountant at the Yarmouth office resolved the problem, and Julio was able to pay his rent arrears.

September was an evil month. The wind was angry, the sea was angry, Vernon and Ricky were perpetually angry. It was wet, cold and miserable; there were complications with the drilling, and everyone was thoroughly fed up.

Fierce gales lashed the rig day after day, and though the supply boats had enormously powerful engines, they were no match for the seas, and time and again the fuel, cement and water lines were torn away when the boat was unable to maintain its position when discharging. One giant wave swept over the *Smit Lloyd 6* and washed a huge food-container over the side as if it was a cardboard box; before it could be recovered, it sank and food worth hundreds of pounds was lost. A sling snapped as the crane was moving a stack of pipe on deck, and Antonio, one of the Spanish roustabouts, was badly crushed and lay in my cabin pumped full of morphine until the wind moderated sufficiently to enable a helicopter to fly out for him.

Special equipment was needed urgently to tackle serious problems on the rig floor. It was too rough to bring it out by boat and too bulky to carry inside a helicopter, but the resourceful pilots risked their necks and brought it to the rig slung underneath. Very gingerly, it was lowered onto the helideck and un-hooked, and the helicopter returned to Yarmouth.

All the rigs in the North Sea were being battered by the gales, and a rig a few miles away had a terrible tragedy. During a lull, a supply boat tied up against the rig to offload, but a gale sprang up again, whipping up enormous seas that stretched the thick nylon mooring-ropes almost to snapping-point, and the boat was being pounded against the rig legs. Grabbing an axe, the mate ran down from the wheelhouse to chop through the ropes, but as he reached the deck a steel mooring-bollard broke loose, and with the tremendous tension on the rope it hurtled across the deck, striking him on the head and killing him instantly. The rig radioed to ask if we had a doctor, but there was little anyone could have done for a man whose head had been smashed to pulp.

The single-engined Whirlwind helicopter on charter to Phillips was not allowed to fly if the wind strength was particularly violent, and it played havoc with the crew-change schedules. When it did arrive, there was almost a stampede to climb in and get as far away as possible from the rig, the sea, the noise, the cold, the hard work and the relentless gale-force wind that flung stinging salt spray across the deck and made every metal ladder and companionway as slippery as an ice-rink. The one consolation for those left behind was that the helicopter brought a mail sack with letters for the lucky ones, newspapers, magazines and, best of all, a fresh supply of cine films.

A few minutes after I had dumped the mail in the recreation-room, Larry appeared, his face wreathed in smiles.

'Ah've got somethin' for ya,' he said, thrusting a package into my hands. 'It's from me and Rose-Ann.' I was taken aback and stared at it for a moment. 'Go on, open it!' he said eagerly. 'It's what you bin' wantin'.'

I tore the wrapper off and pulled out a copy of an American book - *Boatbuilding in Your Own Backyard* by Sam Rable. Inside the front cover, in Rose-Ann's handwriting, was the inscription 'To Bob the radio man, from Larry and Rose-Ann.' I stood holding it, absolutely lost for words, and Larry's face clouded.

'Gee, don't you like it, Bob?' he asked anxiously.

'Hell, Larry, I can't thank you enough,' I said, fighting back a lump in my throat.

It was a superb book, and one I had hankered after for years. He had heard me talk about it many times and in his slow, laborious writing must have added a note to one of his letters to Rose-Ann. Using the precious dollars needed for her wedding, she would have sent for it by mail order and in a remote shack in the Appalachian Mountains had carefully wrapped it and sent it to someone on the other side of the world who to her was just a name. I was deeply moved.

Her letter to Larry was all about the wedding dress her Ma had made from the length of white silk the travelling salesman

had let her have at a reduced price. She was going to have flowers pinned in her hair, and Aunt Belle said she would send for a
pair of white shoes for her out of the Sears Roebuck catalogue.
Rose-Ann wanted May-Lou to be bridesmaid, but the preacher
had said no, . . . 'on account of what she'd been doin' with Uncle
Jeb she weren't a maid no more.' I asked Pa what he meant, but
he went all red and said to ask Ma, and Ma said it was somethin'
a woman got to know about when she was married.' She enclosed a photograph of herself sitting on a horsedrawn 'buggy',
and she looked very young and attractive.

'Goddam!' exclaimed Larry, planting a big kiss on the photograph. 'If she ain't the purtiest woman in the whole goddam
world! When mah contract's finished on this sonofabitch, ah
ain't never gonna move from home no more, you betcha!'

Graham's rig had been scheduled to move to a location off
the Danish coast as soon as the weather improved, and it would
be out of range of my transmitters, so the evening before I was
due to go ashore for my break I called the rig to tell him about
the book Larry had given me. There was a lot of atmospheric
interference, and reception was poor, but eventually I raised the
radio-operator and asked if Graham was available. His transmission was very broken and I could make out only a few words.

'I didn't get you, Harry,' I said into the microphone. 'Can
you try again?'

The radio crackled again, and I heard the words, 'Graham's
gone.' I pressed the transmission button again.

'Hello, Harry. I think I got that. Where's Graham gone? Has
he quit?'

There was more crackling and I pressed my headphones
to my ears to cut out the noise of the rig. Harry's voice was
very weak, but the words bored through my head like a drill.

'Hello, Hewett "A". Sorry to have to tell you that Graham has
gone over the side. The crane he was operating fell into the sea.
It's too rough for divers to go down, but there's no chance that
he's survived.'

I sat, numbed with shock. Harry's voice broke through again:
'Hello, Bob. Did you get that?'

In a daze I picked up the microphone. 'Yes, I got it, Harry. Thanks.'

I went to my cabin and lay on my bunk. Suddenly I hated rigs. It was a cruel, savage way of life, abandoned by a compassionate God and where only the most iron-hearted could survive. Why should a cheerful, talented, hard-working man in his early twenties, who enjoyed life to the full and had great ambitions, be snuffed out like a candle? What made people cling to ludicrous religious beliefs that watching over them was a Saviour, a great protector, when all around them they were being let down? What possible compensation was there for the misery that Graham's parents would endure? My mind was in a turmoil, and all night I tossed and turned in the grip of a dreadful nightmare, in which Graham climbed into the crane cab and I reached forward to stop him but could not move or speak. I watched, helpless, as the crane slowly toppled over the side of the rig into the sea, and I saw Graham float to the surface. He smiled up at me, then sank beneath the waves. It was horrible, and I awoke soaked in sweat.

I went home vowing never to return to the Hewett or any other rig. Very depressed, I neglected my family and, taking a tent, some food, a sleeping-bag and a rucksack, went into the mountains and camped by the side of Burnmoor Tarn, high on the fells above Wasdale.

During the week, I climbed the peaks of Scafell, Scafell Pike, Yewbarrow and Great Gable and in the evening sat outside the tent in the perfect stillness and watched the setting sun sink below the horizon. By the light of a flickering candle I worked my way through a wonderfully inspiring book about how sixteen Estonian refugees, including four young children and an old lady, crammed into a leaky thirty-seven-foot yacht and sailed it from Sweden to the United States to escape from the Russians. It was a story of perseverance and hope and, full of admiration for their courage, I felt a lot better.

I returned home and went to visit an old retired sea captain who had taught me navigation. If anyone had been to hell and back, it was he. During the war he had been torpedoed no fewer than four times and survived with hardly a scratch. The first time he was torpedoed, the German U-boat surfaced and machine-gunned the survivors in the lifeboats, killing everyone but himself and the ship's cook. They clung to wreckage for days before they were rescued by a Royal Navy destroyer, but the cook died as he was being lifted aboard. The captain was a very philosophical man, who had been able to forgive, though not forget. He had no time for religion and firmly believed that from the day we are born the method and time of each individual's departure from life are pre-determined.

'For those poor sods gunned down by the Krauts, their number was up; but mine wasn't and I survived. When my other ships were torpedoed, it was tough that as we drifted in the lifeboats the men who perished had to endure a long, lingering death from hunger and thirst or the intense cold, but that's what fate had planned for them, and there's bugger-all anyone can do about it. It's no use those Bible-punchers spouting about the love of God and singing, "Oh hear us, when we cry to thee for those in peril on the sea." Nobody helped those poor bastards lying in the boat with their eyes popping out and their tongues black and swollen through lack of water. It's the same with your mate on the rig. He died because his number was up, and there's no use you getting yourself wound up and looking for someone to pin the blame on. Will Shakespeare hit the nail right on the head when he wrote, "All the world's a stage and all the men and women merely players. They have their exits and their entrances." The only change I'd make to that is to say that all the world's a tragic stage!'

Talking to my old friend was just the tonic I needed and, tearing up my letter of resignation, I returned to the rig with renewed strength.

Scotch Jimmy was not on the helicopter when the roustabout crew-change arrived, and the following day the Rimrock office in Yarmouth called to say that it was unlikely we would see him again, as he was in gaol, charged with assault. I felt sure that his gift of the gab would soon unlock the cell door, and whenever a boat or a helicopter arrived I expected to hear his sonorous voice declaring that the end of the world was nigh, and making sure, with surreptitious gulps from a half bottle, that when it came he would be well fortified. Throughout the week boats and helicopters came and went but there was no sign of Jimmy, and fearing the worst I put a call through Humber Radio to a friend in the Glasgow police.

It was a sorrowful story. Instead of getting boozed out of his mind as usual for the first two days of his shore leave, Jimmy had gone straight home and discovered his wife in the throes of vigorous love-making with the rent-collector. Ever the one to forgive sinners who strayed from the path of righteousness, Jimmy launched into an impassioned sermon on the Ten Commandments, but made the mistake of emphasizing the finer points by beating his wife and her lover over the head with a whisky bottle and putting them both in hospital. Jimmy conducted his own defence in the local sheriff's court, but he was no Perry Mason and, totally befuddled by a barrage of quotations and parables from the Scriptures, the sheriff remanded him in custody pending trial. The general opinion was that, taking his record into consideration, he was almost certain to go to gaol again. The evening discussions in the radio-room were never quite the same without the loquacious Scotsman.

Among the Americans on the rig the conversation was domi-nated by the approaching general election in the United States. John F. Kennedy's successor, President Lyndon Johnson, had announced that he would not stand for re-election, and a 'no punches pulled' election battle was mounting between the Democrat Hubert Humphrey and the wily Californian lawyer Richard Nixon, for the Republicans. One of the penalties the Americans had to pay for earning tax-free salaries overseas was losing their right to vote but, undeterred, the American commu-nity in Yarmouth was planning to have a party in November and hold their own election.

For the first time in months there was no letter for Larry from Rose-Ann when the mail arrived, and he was so agitated that he could not eat or sleep. He pulled the letter rack off the recrea-tion-room wall and turned it upside down to make sure no letter was wedged in the pigeonholes.

'You're wasting your time, Larry,' I said when he appeared at the radio-room door for the tenth time, asking if he could look in the mail sack. 'There isn't a letter for you. Nobody opens the mail sack but me, and I always turn it inside out to make sure nothing is left in it. Maybe mail in the States has been held up because of the election.'

Realizing that this was a possibility, he perked up and helped me to check the level of the electrolyte in the emergency batter-ies outside the radio-room.

'If Rose-Ann's wearing a white wedding dress, what are you going to wear for the big day?' I asked, as he unscrewed the plugs off the tops of the batteries and I squirted in distilled wa-ter. Lifting his hard hat, he scratched his head absentmindedly.

'Jeez, ah ain't never thought about that. When ah was in Houston, Texas, ah got me some Mexican boots and a fancy shirt and pants and a stetson, but the preacher don't like no dude outfits.'

'That doesn't sound like a dude outfit to me,' I said. 'I'm sure Rose-Ann will think you look great. Anyway, there's only a month to go to the wedding, and you've got another tour to do on the rig before you fly home. You'll have no time for shopping.'

With a long sigh, he sat on the battery box. 'Hell, ah'm scared! All the folks at the weddin' will be listenin' and watchin'.'

'There's nothing to it,' I assured him. 'I've been through it all myself. Get plenty of home-made whisky inside you the night before and you'll hardly remember a thing about it.'

He laughed gleefully. 'Ah guess that ain't such a bad idea. Rose-Ann's Pa makes liquor that's got a kick like a mule!'

After completing the work on the batteries, we picked up mugs of coffee and doughnuts from the galley, and in the quiet of my cabin we wrote a long letter to Rose-Ann, telling her what he had been doing on the rig and that he was planning to fly home early in November. He finished it, as he always did, by laboriously scrawling, 'from your lovin Larry' across the bottom of the page; and hammering the stamp onto the envelope with his fist, he departed to his bunk a happy man.

The *Smit Lloyd 6* arrived the following evening with cement and casing, and I chatted for a while with the skipper over the radio.

'By the way,' he said, as he was about to return to supervise the offloading. 'We have a mail sack on board for you. If you'll let the crane-operator know, I'll get one of the crew to tie it on the crane hook.'

Sure enough, there was a letter for Larry, and as he was on the midday to midnight tour I put it in my cabin and sent him a message to say a letter had arrived. The boat was still offloading at midnight, and I was in the radio-room when Larry hurried in, eager for me to read his letter to him. His reading had improved enormously over the weeks, but he found it easier if I read the letter first, then he would go over it again later. I picked up the envelope. Although it had Rose-Ann's address on the back, the handwriting was unfamiliar. Tearing it open, I pulled out the letter and read:

Dear Larry,

I shore don't know how to begin to tell you this and you're gonna have to be very brave.

Last Sunday, one of them Jason boys got drunk and stole a truck and the cops was chasin him just as folks was comin out of the meetin house. The truck skidded in the rain and afore we knowed anythin it piled into Rose-Ann and killed her dead. We buried her yesterday and folks from all over come and brought flowers and paid their respects. Me and her Pa asked the preacher how could God be so cruel as to take our little girl and he said to consider it this way. If you was agoin to pick a flower in the garden, you shore would pick the most beautiful one and that's what God did. We are eatin our hearts out for you Larry, but we know that God will take good care of Rose-Ann.

May the Good Lord give you strength,

Jane Fleming.


Larry's excited voice broke through my paralysed brain. 'Come on, Bob. What's she say? What's she say?' I tried to speak, but shock gripped the muscles in my throat, and words would not come.

'Quit foolin', Bob,' cried Larry impatiently. 'What's she say?'

'Sit down, Larry,' I said hoarsely. 'I've got bad news.' He looked at me in alarm. 'She ain't sick, is she?' 'Larry,' I said, as gently as I could, 'Larry, Rose-Ann has been involved in an accident.'

His eyes widened. 'Wadya mean, accident?'

I avoided his face and stared at the floor. 'Larry,' I said quietly, 'she's been killed by a drunk driver.'

He looked at me with a puzzled expression, as if he could not understand what I was saying, but when the realization dawned he launched himself at me and gripped the front of my shirt.

'You lyin' sonofabitch!' he sobbed. 'Whatya tellin' me that bullshit for? Rose-Ann ain't dead! She just writ me a letter. You ain't no buddy of mine tellin' me that Rose-Ann's dead. Ah ain't never give you no hurt.'

'Take it easy, Larry,' I said, trying to stay calm. 'I'll read you the letter.'

Releasing his grip on my shirt, he sank onto a chair, and tears streamed down his face. When I finished reading, he slid to the floor with his hands clasped round his head, whimpering, 'Rose-Ann, Rose-Ann. Where are you, darlin'? You ain't dead, ah know you ain't dead!'

He was in a bad state of shock, and several hours passed before I was able to persuade him to lie on the spare bunk in my cabin. When I told Ricky what had happened, he was remarkably sympathetic and understanding and excused Larry from work. Two days later he went ashore with the crew-change, but he was a physical and mental wreck. I was deeply worried that he would head for the nearest bar and drink himself senseless, and when I was relieved a few days later I went to Yarmouth in search of him. I combed all the pubs, dancehalls and drinking-dives I could think of, but he had disappeared.

I tried the hospital, the police and the Petroleum Club without success and was about to give up and start the long journey home to the Lake District when I met the captain of one of the supply boats who had seen Larry the night before in a hotel near the harbour. It was a hotel notorious for its shady activities, and as soon as I walked through the door I was offered the 'menu', which included dishes like Julie, Gloria or Emma at £20 for the night, or £100 if I wanted to be really kinky and try the speciality of the house, with flogging included. It was early afternoon, the dining-room and bars were virtually empty and there was no sign of Larry. Near the main door a grubby sign pointed to the 'cellar bar', and I descended a flight of concrete steps to a badly lit dungeon reeking of stale beer and cigarette smoke. A jukebox was blaring out the latest Beatles hit, and on a stool at the bar a beefy American driller from one of the International Drilling Company rigs was passionately entwined round a plump blonde. I could vaguely make out a few other men, drinking and playing cards, and as my eyes became accustomed to the gloom I saw him, slumped across a table.

'Larry,' I said, shaking his shoulder gently, 'come on, let's get out of here.'

He grunted, raised himself on one elbow and peered at me with bloodshot eyes.

'Well, I'll be doggone,' he slurred, 'if it ain't mah 'ole Limey buddy the radio man. Hey, bartender,' he shouted, 'fetch mah buddy a drink. Make it a Scotch.'

'I don't want a drink, Larry,' I said. 'Let's get out of this dump and sober you up.'

One of the house girls sidled up and put her arms round him.

'You don't want to go yet, do you, honey?' she simpered. 'Let Julie take care of you, and when we've had a few drinks maybe we'll go and lie down for a while.'

'Can't you see he's in a bad way?' I said to the girl. 'Help me to get him to a taxi.'

'Why don't you piss off?' she snarled. 'Larry's OK with me, aren't you, honey?'

Larry tried to stand, but he fell back giggling into the girl's arms.

'Yeah,' he sniggered. 'Ah'm just fine an dandy with mah li'l ole Julie. Why don't ya piss off, radio man?'

He slumped forward over the table again and I shook him vigorously.

'Larry, come on. I'll get you to a taxi.'

'Ah don't want no taxi,' mumbled Larry. 'Piss off, radio man.'

I felt a tap on my shoulder and turned to find the driller towering over me.

'You heard what Larry said. He's OK with us, so beat it!' The jukebox had died and there was a deathly hush in the room as the drinkers put down their glasses and waited for my reaction. I looked at Larry. He was weeping uncontrollably, and the girl had opened her blouse and pressed his head against her breast. Her free hand was working its way down his back towards the wallet sticking out of his hip pocket.

'Beat it!' repeated the driller.

Larry did not return to the rig, and a few weeks later he was flown back to the USA, a broken man.

November arrived in a bellicose mood, with a strong wind heaping the sea into a rolling, tumbling mass that spent its malevolent strength against the rig with a shuddering crash. Vicious squalls of rain and sleet continually lashed the crews working on deck, and woollen clothing and gloves under thick oilskin suits did little to keep out the biting cold. When the men finished a twelve-hour tour, their faces were grey with fatigue and their fingers and toes so stiff they were hardly able to move. At the end of the first week the wind dropped, and for a day or so there was a threatening stillness, with thick black clouds hanging motionless in the sky. The sea stirred uneasily, as if anticipating something evil in the air, and the birds resting on the helideck sensed it too. Starlings that had been roosting there for weeks deserted us, and even the voracious gulls left titbits untouched.

The first week of my two-week tour on the rig always dragged by painfully slowly. It was like driving an old car up a steep hill, taking ages to reach the top but once over the summit a rapid run downhill to the bottom. On the Monday morning of my second week on the rig, I was busily polishing the radio-room floor, with a light heart. For once it had been an uneventful tour. Snug inside the radio-room, I was one of the fortunate ones who could shelter from the weather. There had not been a solitary accident all week, not as much as a cut finger. The radio equipment was in perfect order, and I had managed to have a long conversation on Sunday with a sea-captain friend as his ship was leaving the Thames estuary, bound for India. An offer I had put in for a fishing-boat which I planned to convert to sail had been accepted, and, nicest of all, in a few days I would be at home. Life was very pleasant.

About midmorning I heard a ship calling the rig, and when I answered I found to my surprise that it was the trawler *Hector Gannet*, which in 1965 had been the supply boat for Mr Cap, the first rig to drill in the North Sea. More recently it had been the standby boat on the scene during the Ocean Prince disaster. The skipper said that he had a pallet of equipment for the Hewett and had been ordered to relieve the *Smit Lloyd 6* as the rig standby boat. Interested to see the boat that had been involved in gas and oil exploration in the North Sea virtually from the day it started, I went on deck and watched it come alongside. A converted stern trawler with forward wheelhouse and flat afterdeck, it was not as large or as manoeuvrable as the Dutch and German boats, but it was built for trawling in any weather and was ideal for rig standby work.

In the evening, when the boat lay at anchor, I chatted to the skipper, Lionel Tomlinson, an ex-trawlerman from North Shields. I was curious about the name *Hector Gannet,* and he explained that it was one of two stern trawlers, *Hector Gull* and *Hector Gannet,* once owned by the London-based Hector Trawling Company, which was involved in whaling. He also told me that the mate on board with him, Harry Goodison, had been the skipper of the *Hector Gannet* at the time of the Ocean Prince epic, and later, when Harry came on watch, I had a long talk with him about it.

'Couldn't bloody get near enough to do anything for 'em,' he said in a gruff Geordie accent. 'We were hove to in heavy seas and just had to watch, helpless, as the rig fell to bits. I wouldn't like to go through anything like that again.'

Tuesday morning was bleak, bitterly cold and so dark at ten in the morning that I thought the radio-room clock had gone haywire and I had leapt out of my bunk three hours early. The *Hector Gannet* was rolling heavily at anchor, but when I spoke to Harry he said that the only thing bothering the crew was that they were running out of reading-material. I said I would have a scour round the mess-room and drop a parcel on their deck when the weather improved enough for the boat to come alongside. Just before lunch a helicopter arrived, with Vernon, a bulging mail sack, a Phillips geologist and several service company engineers.

For most of the crew the arrival of the mail sack was eagerly awaited; but for the Americans with relatives serving in Vietnam it was an ill omen, and the familiar airmail envelopes would often lie in the letter rack for several days before whoever they were addressed to plucked up enough courage to open them. The cheerless headlines in the tabloid newspapers did little to ease their worries, and the situation in Vietnam seemed to worsen with every edition.

There were two letters for me. One contained details and photographs of the fishing-boat I had bought, and the other was from an agency offering me a job as mate on a large schooner chartering in the West Indies. It had been the same throughout my seagoing career. The only time I was ever offered a dream of a job like that was when I had signed a contract to work for someone else. With the usual American impatience, they wanted a decision by return of post.

Apart from the porn films for which there was an insatiable demand, the most popular films shown on the rig were Westerns, and in the mail sack there were no fewer than three. Among the Americans on the rig there were at least two full-blooded Apache Indians, and I often wondered what ran through their minds as they sat impassively while the US cavalry mowed down their ancestors with fiendish abandon and the men watching the film whooped and roared with delight. The only time there was any reaction was during a film about General Custer's last stand at the battle of Little Bighorn. For once, the cavalry were on the losing side, and as the troopers dropped with arrows in their backs, there were grunts of approval from the Apaches. At the height of a particularly gory scene, when a trooper's scalp was being lifted off like the top of a boiled egg, a young roustabout, waving his arms about too enthusiastically, clouted one of the Apaches across the head. The big Indian reached out and caught the roustabout by his hair.

'Be careful, feller,' he growled. 'You saw what happened to that white man!'

The men coming off tour at midnight on Tuesday set up the projector to show one of the films, but it was a new breed of Western with Clint Eastwood, cigarette dangling from the corner of his mouth, eyes narrowed to slits and so laid back that when he struck a match to light his cigarette there was time to nip to the galley for a cup of coffee and be back before he inhaled the first puff.

Bored to distraction, I went to my bunk and read through a book of American humour. There was a story about a Texas oil-man who was lying in a hospital bed and, feeling randy, said to the nurse, 'Say, ain't you gonna give me a bed bath?'

'OK,' said the nurse, and went away to get a bowl of water. The Texan was so rampant with excitement that it looked as if the blankets were airborne, but the nurse was experienced in handling this sort of situation.

'Right,' she said, 'I'll wash up as far as possible, then I'll wash down as far as possible. Then I'll give you the soap and sponge and you can wash old possible yourself.'

Wednesday showed every sign of being a quiet, uneventful day. There were no helicopters or ships due, and Vernon, who had been up all night working with the drilling crew, had flopped onto his bunk. After breakfast I talked for a while with Harry on the *Hector Gannet*. He was very bitter about the Government's lack of support for the British trawler fleet.

'There's hardly a bloody boat left in North Shields now,' he complained, 'and it's the same in Hull and Grimsby and any of the big ports you can mention. It's a waste of time talking to politicians. When I was at home, I heard one daft bugger on the wireless say that Yarmouth was the country's leading herring-fishing port. It just shows how out of touch those idiots are with what's happening in the fishing industry. Yarmouth was finished as a fishing port over twenty years ago.'

In the afternoon there was a minor emergency in the galley when a steward knocked a pan of boiling water off a cooker and scalded his foot. It was not serious, and as soon as I had put a dressing on it, Archie the chef made the lad go back to work again.

Thursday was wet and miserable, and the shipping forecast at 6 p.m. warned of a south-easterly gale force 8, perhaps force 9 at times, for the southern North Sea. The wind-speed indicator was already hovering around 35 mph (force 7), and the sea was streaked with white spume. When I chatted to Lionel Tomlinson on the *Hector Gannet,* he reported a rising sea but no problems.

There was another midnight film show but John Wayne's slow drawl was lost in the howling of the wind sweeping around the rig, and taking a mug of coffee to my cabin, I clamped head-phones over my ears and plugged in to my portable cassette-player. Lying with eyes closed, listening to loud music through headphones, was the only way to escape from the realization of being perched on stilts only a few feet beyond the grasp of one of the wildest and most notorious of seas in the world. The sound track from the film *Love story* was so soothing that I fell asleep listening to it.

At 7 a.m. on Friday I woke feeling refreshed and, after my usual breakfast of coffee and toast, ventured on deck for a breath of air. The rig was being blasted from the south-east by a full gale, and it was blowing straight at the men working on the drill floor. They were pulling out of the hole and, as the sections of pipe were unscrewed, thick mud spurted out like a fountain and the men were plastered from head to foot. They had been work-ing since midnight, exposed to the full force of the wind and spray, and there were still five hours of their tour to go. For a lucky few it was crew-change day, and at 8.30 a.m. Bristows radioed for a weather report. The windspeed-indicator was swinging between 45 and 55 mph (force 8 to 9), but at 9.30 a.m. Bristows called again to say that the helicopter was on its way. When I spoke to Lionel Tomlinson on the *Hector Gannet,* he said they'd had a rough night and that some of the seas were at least fifteen feet high and it was too dangerous for his crew to work on deck.

'You're better off on the rig in weather like this!' he joked. 'All that lovely grub and films every night. It sounds like a lux-ury hotel up there.'

The helicopter landed at 9.50 a.m., with Ron Gower, his roustabout crew and two service engineers, and within minutes it was airborne again, heading back to Yarmouth. It was the cus-tom for the newly arrived crew to have a cup of coffee before changing for work, and on my way through to the recreation-room with mail I had a quick word with Ron. The appetising aroma from the galley made me feel hungry, and I coaxed the chef to make me a bacon sandwich.

'Now sod off back to the radio-room,' he bantered, handing over the sandwich on a plate. 'This is a kitchen, not a bleeding hotdog stand.'

'OK, oh mighty chef,' I mocked. 'What gastronomic extravagance are we having today then?'

'*Entrecôte chasseur*, or fancy sirloin steak to peasants like you!' he retorted.

'That sounds great,' I said. 'Isn't it marvellous what you can buy in a can these days?'

Suggesting to Archie, who had worked in the best hotels in Europe, that he prepared his meals by opening cans was like lighting a match under a keg of gunpowder. I escaped to the radio-room before he exploded!

At 10.15 a.m. Vernon appeared, swathed in oilskins and dripping with water.

'Jesus, what a day!' he exclaimed. 'Those guys on the drill floor are almost frozen to the balls.'

He flung his oilskins into a corner, went to the galley and brought back a cup of coffee. We were sitting talking when a tremendous explosion rocked the rig, followed by an ear-splitting roar, as if we were suddenly caught in the grip of an immense storm. The force of the explosion threw me off my chair but Vernon managed to keep his feet and, throwing his coffee cup to the floor, raced outside. Alarm sirens were sounding all over the rig, and I could hear shouts on deck. Within a few minutes, Vernon was back.

'She's blowed out!' he gasped. 'Get that standby boat in right away!'

For the drilling crew on the Hewett "A", the morning of Friday 15 November was like any other morning, except that, as for the past week, it was wet, windy and cold. Since midnight they had been pulling out of the hole, and it had been the same monotonous routine. The massive pulley block in the derrick heaved a ninety-foot section of drill pipe out of the hole, and it was unscrewed and swung to the side of the derrick. On the monkeyboard, high above the drill floor, the derrick man guided the pipe into a rack. The pulley block descended, another ninety-foot section was heaved up and the process was repeated. Each time a section of pipe was unscrewed, black mud spurted out and showered the men with a slimy mess that dripped from their hard hats and oilskins. Motors roared, chains clanked, metal squealed against metal, mud and sweat mingled on the men's faces and ran down their necks, soaking sweaters and shirts. And it went on . . . hour after weary hour.

For most of the time the windboards fastened round three sides of the base of the derrick gave some protection from gales, but on that Friday morning the wind was blowing hard from the south-east directly onto the drill floor. It penetrated the thickest clothing like a knife, and to add to the discomfort the mud spewing out of the drill pipe was caught by the wind and sprayed everywhere, as though from an aerosol can. Within just minutes of starting work, everyone was thoroughly wet, cold and miserable, but on the rig there was no wet time, dirt money, hardship allowance or any of the other perks enjoyed by workers in outdoor occupations on shore. At six o'clock it had stopped raining, but the gale was blowing as hard as ever. Taking it in turns, the crew had a break for half an hour and snatched a meal and a mug or two of hot coffee.

For the crane-operator and the roustabouts it had been an equally dreary night, and they were worn out. Vernon had left instructions that the stacks of drill pipe had to be moved from one side of the rig to the other, and they had been working on deck in the teeth of the gale. Salt spray clung to every pipe, every wire sling and metal stanchion, and the greasy deck was like an ice-rink. They had been buffeted and pummelled by the fierce wind, drenched with freezing water, had torn their oilskins on jagged metal, been sworn at by a bad-tempered toolpusher and were expected to work in conditions that were sub-human.

After the 6 a.m. break, two young roustabouts who had been on the rig less than a week refused to go back to work, saying it was slave labour and they were going ashore on the crew-change helicopter due that morning. Joe, the crane-operator, said nothing. Pulling on his oilskins, he jammed his hard hat firmly on his head and went out to his crane. He had heard it all before. The young kids were taken in by the talk of big money to be made on the rigs, but when they came out and discovered how hard they had to work for it, they quit within a few days. Joe started the crane and swung the jib round towards the group of roustabouts huddled in the shelter of a stack of pipe.

'It's going to be tough on them,' he muttered to himself, 'but the poor buggers will just have to work that bit harder until two more hopefuls are hired.'

In the engine-room Yorkie, the assistant rig mechanic, was wiping spilt oil off the big diesel engines with a piece of rag and singing at the top of his voice. No one could hear him - in fact, he could not hear himself, for the four eight-cylinder diesel engines, all running together, made a lot of noise. But he could feel the words of the song in his throat. A clock on the wall said 8 a.m., and today he was going ashore. The American rig mechanic was due back after going home to attend the funeral of his son killed in Vietnam and taking a few weeks vacation. He liked the American, and they got on well together, although it irritated him the way some of the Americans had a sneering opinion of British mechanics. Yorkie had served his apprenticeship with one of the world's leading marine engineering firms, and yet time and again when he was expecting to be promoted to rig mechanic the American companies had given it to one of their own men. He was always quick to tell a tale about a time when he worked on a rig and had to work as assistant to an American mechanic who openly boasted, 'A coupla years ago ah couldn't even spell "engineer", and now ah am one!' With a bit of luck, there might be good news in the post when he reached home. Recently he had filled in an application for a job as rig mechanic with British Petroleum on their rig Sea Quest.

In his cabin that doubled as an office, a store-room, and a general depository for miscellaneous junk, Archie, the chef, was flicking through his menu file. 'File' was probably too grand a title for a battered shoebox containing hundreds of bits of paper with recipes scribbled on them and pages stolen from cookery books, but it had been with him since his catering college days, and he could not bear to throw it away. The file had been added to in Paris, Geneva, London and Rome, where he had prepared meals for the crowned heads of Europe, film stars and assorted mega-rich. His photo album, another crumbling shoebox, was full of pictures of himself standing beside a gigantic cake or an elaborate meal or shaking hands with people whose names made world headlines.

'What a bloody come-down,' he muttered to himself, as he gazed wistfully at a menu he had prepared for Aristotle Onassis aboard his yacht *Christina*, 'from feeding the great to feeding glorified building-site workers who don't know a croissant from a quiche.'

The story of his fall from fame was an anguished tale of passion which sounded suspiciously as if it had been poached from Shakespeare's most famous love story. He had fallen in love with the daughter of one of Europe's royal families, and though the princess felt the same way about Archie, her parents, not surprisingly, disapproved and sent her to school in Switzerland. Archie followed, but the princess was moved to another ivory tower, and through the aristocratic grapevine the girl's mother had Archie declared *persona non grata* in upper-class kitchens throughout Europe. Unable to get work, Archie went to sea as a ship's cook until he joined the rig catering company. It was a touching story, but the true reason for his demise was in the clink of glass each time he reached under his bunk and pulled out another of his shoebox files.

Glancing at his watch, Archie saw that it was 8.30 a.m. There was a new steward due on the crew-change helicopter, and he was not sure which work rota to put him on. First, though, there was the midday meal to decide on. He flicked thoughtfully through the menu file. There always had to be several choices at each meal, and with the Americans meat was top priority, and it had to be beef. He shuddered at the memory of the first rig he had been sent to, where he had prepared a leg of lamb. The Americans sniffed at it suspiciously and poked it with their forks.

'Say, what's this heah meat?' demanded the big Texan tool-pusher.

'It's a superb leg of lamb,' he had replied, anticipating the praise to come.

'Lamb!' roared the Texan. 'Lamb! Texans don't eat that shit. Ah wants nothing but beef steak on this heah rig.' And picking up the tray of meat off the servery, he carried it on deck and threw the lot into the sea.

Archie picked a piece of paper out of his file, and studied it briefly. 'Sirloin steak, just the job,' he said to himself. 'I'll dress it up and give it its French name, *Entrecôte chasseur*. That'll give 'em something to chew on!' Chuckling at his own joke, he went out to the galley.

Just after 9.30 a.m. the Hewett crew-change helicopter lifted off from Yarmouth. On board were Ron Gower, the crane-operator, half a dozen roustabouts, two service company engineers and a catering steward. It was a bumpy flight, with the helicopter lurching about the sky in the strong squalls, and as it descended towards the rig the men could see the *Hector Gannet* rolling and pitching at anchor, and large waves crashing against the side of the rig, throwing plumes of spray high into the air. Within the space of a few minutes the helicopter landed, exchanged crews and was airborne again. The pilots always made it look so easy, yet they had to manoeuvre a fully laden helicopter onto a landing-pad forty feet in diameter, a hundred feet above the sea, in a 50 mph wind. A magnificent feat.

Ron Gower and his crew made their way to the galley for coffee before starting work at midday. The service engineers disappeared in search of a cabin, and Archie had the new steward scrubbing passageways almost before he had time to change into working clothes.

On the drill floor the drilling crew were nearing the end of the long pull out of the hole. There were only a few more sections left to pull before the drill bit was reached, and the men worked with renewed energy. A section of pipe was lifted up by the travelling block, and as the crew unscrewed the section a column of mud shot up the derrick. At first they thought it was nothing more than pressure in the pipe, but as it rose higher and higher the American derrickman realized they were in trouble.

'Get the hell outa there!' he yelled down at the men on the drill floor. 'She's blowing out!'

Grabbing the 'dead line' - an emergency rope hanging permanently in the derrick to slide down to the drill floor - he forgot that his gloves were covered in mud and, unable to grip the rope to control his descent, landed heavily on the drill floor, breaking both ankles. The crew half ran and half dragged him across the deck and had reached the accommodation when there was a tremendous bang, and an enormous column of mud and gas shot up the derrick. Alarm sirens went off all over the rig, and Vernon rushed on deck.

'She's blowed Vernon, she's blowed,' shouted the derrickman. 'We gotta get off this mother, and quick!'

Vernon weighed the situation up quickly. 'Get everyone on deck, wearing life-jackets,' he bellowed to the driller. 'Get the roustabouts to check there's no one left in the cabins.'

Within ten minutes drilling crews, roustabouts, caterers, service engineers, rig mechanics, electricians and J. C. Youngblood, the Phillips company man, were assembled on deck. The noise from the blow-out blasted their eardrums, and there was imminent danger of an explosion from an accidental spark; but though there were many anxious faces, there was no panic.

'The standby boat's coming in,' yelled Vernon above the roar. 'Get the scramblin' nets over.'

The *Hector Gannet* was approaching on the east side, and the men rushed to drop the net and climb down it; but the skipper waved to the men to indicate that he would come in on the west side.

The heavy net was thrown down on the west side, and several men climbed down and hung on, waiting for the boat to come in. The *Hector Gannet* was being thrown about in the heavy seas, and as the skipper brought the stern in under the net, a big sea lifted it and slammed the hull down onto the boat landing, breaking off a girder covered in truck tyres which normally acted as a fender. The skipper managed to hold the boat long enough for twelve men to drop onto the deck. The men on the rig saw one man fall heavily and, unable to walk, he was dragged along the deck by the seamen. At the bottom of the net two young roustabouts, frozen with fear and cold, could not let go of the net. 'Jump!' yelled the men on the boat and held out their arms to catch them, but they looped their arms in the net, eyes wide with terror, not daring to move.

Another big sea lifted the stern of the boat, and it dropped again onto the boat landing, but this time, unknown to anyone, the broken girder pierced the hull below the waterline, and the hold began to take in water. The plight of the roustabouts on the net was serious. They were tiring rapidly and were losing their grip. Louis Sylvester, a Canadian driller, and Richard Craig from Whitby, one of the drilling crew, snatched up thin lengths of nylon cord and with great courage climbed down the billowing net and tied them round the waists of the roustabouts to prevent them from falling. Flogged by the gale and soaked with spray from the waves, their fingers were so cold they could hardly tie the knots. Shouting to the men on the rig to heave on the ropes, Louis and Richard helped the men up the fifty-foot net, a rung at a time, and they were pulled to safety.

While this drama was going on, the rest of the men on the rig watched horror-stricken as the *Hector Gannet* moved away, listing heavily, and then rolled onto its side, finally sinking, leaving the men floundering in the water. Inflatable life-rafts from the rig were hurled into the sea, but the strong wind blew them away like kites. A number of men were affected by the gas and, worried that it might be poisonous, Vernon herded everyone into the shelter of the mess-room. It was an agonizing wait as the rig shuddered and rocked, and a cheer went up when the first helicopter arrived. Soon all the crew were lifted to safety. When the last few men were going up to the helicopter, Vernon shouted, 'We forgot the radio man!' and raced back to the radio-room. The men climbed in and closed the door, and the pilot took off, unaware that two men had been left behind.

On the *Hector Gannet* the crew had finished breakfast, and Lionel, the skipper, and Harry, the mate, were discussing the day's programme. The boat was lying at anchor, and rolling and pitching heavily in the gale.

'What do you want the men to do today?' asked Harry.

Lionel peered through the wheelhouse window at the rough sea and glanced up at the black clouds tearing across the sky. After over forty years at sea and most of them in trawlers, it was just another gale to him, but though the working of the ship had to go on whatever the weather, there was nothing to be gained from putting the crew at risk.

'Don't let 'em work on deck,' he replied. 'It's far too wild. Give 'em a bit of cleaning up to do, and maybe polishing the brasswork.'

Harry nodded and went below to the men. Having adjusted the volume control on his VHF radio, Lionel sat back in his chair. A short distance away heavy seas were breaking against the green-painted legs of the Hewett rig and falling back in a seething mass of foam, which was caught by the gale and flung high into the air. Just before 10 a.m. the rig had called to say that a helicopter was due and to stand by, as usual, in case of an emergency; but it had landed and taken off effortlessly, like a herring gull on the top of a ship's mast. Attending to a few odd jobs in the wheelhouse, Lionel wound up the ship's clock and checked it with his watch. It was just coming up to 10.10 a.m. Hanging the clock key back on its hook, he glanced round at the rig and was startled to see a long column of black stuff squirting out of the top of the derrick.

'Hey, Harry,' he shouted to the mate, 'bring the crew up. I bet they've never seen anything like this. The bloody rig's struck oil!'

The men crowded into the wheelhouse and stared as the black fountain rose higher and higher.

'It's just like the film they made about that American feller, Red somebody or other,' said Harry. 'I remember John Wayne was in it, but I'm damned if I can remember the other feller's name. What was it?'

Before anyone could answer, a voice broke through the VHF: '*Hector Gannet, Hector Gannet,* this is the Hewett "A".'

Lionel grabbed the microphone. 'Hello, Hewett "A". This is the *Gannet.*'

'Lionel, we've had a blow-out. We've got to evacuate the rig. Get alongside as quickly as you can!'

'Harry, call all hands,' said Lionel swiftly. 'Get everyone to put a life-jacket on, and heave anchor.' He turned to the radio again. 'Hello, Hewett "A". This is the *Gannet.* We'll be right with you. Should we cut the chain and slip the anchor?'

There was no reply from the Hewett, and he tried again. There was still no answer, and fearing the worst he called the Arpet platform a few miles away and asked if they thought the emergency justified saving time by cutting free from the anchor. 'Get there as quick as possible!' was the reply.

By the time the anchor was released, the engineers had started the engines, and Lionel steered for the rig. As the boat drew near, there was an overpowering smell of gas, and the cook dashed into the galley to turn off all his stoves.

'No smoking, lads!' Lionel called to the crew. 'And for Christ's sake don't strike any matches.'

The men on the Hewett had heaved scrambling-nets down the east side of the rig and were preparing to climb down, but on that side the seas were breaking heavily and the boat would have been smashed to pieces. Lionel waved to the men to go to the other side of the rig, which was more sheltered from the gale. A scrambling-net on the west side was heaved over, and as the men clung on grimly, Lionel attempted to manoeuvre the *Hector Gannet* stern first under it. The huge swell lifted the boat like a cork and slammed it against the rig, but Lionel managed to keep the stern into the rig long enough for twelve men to drop from the net onto the deck. One man landed heavily and broke his leg, but the seamen dragged him to a cabin and into a bunk. Twice the swell lifted the boat and smashed it against the rig, and the second time as it crashed down the stern landed on a broken girder that tore a hole in the hull.

None of the crew was aware of it until, trying to get the boat back under the net to two men who had been left behind, Lionel noticed that it was listing slightly to starboard.

'Harry,' he shouted to the mate. 'There's two kids hanging frozen on the nets. We've got to get 'em off, but I've a feeling we've been holed. Nip aft will you, and check the hold.'

Two young roustabouts clinging to the bottom of the net were frozen with cold and yelling that they could not hold on much longer.

Lionel was about to put the engines astern to position the boat under the net when Harry rushed into the wheelhouse. 'Lionel, there's a hole in the hull under the stern, and water's rising rapidly in the hold.'

'Set the crew to take all the mattresses off the bunks and see if they can stuff them in the hole,' rapped Lionel. 'I'll have to pull away and bring her head into the wind to steady her.'

'What about the kids?' shouted Harry over his shoulder as he raced away to get the crew.

'We'll try and get back for them,' called Lionel, 'but we'll have to plug that hole first.'

By now, the *Hector Gannet* was listing heavily, and as it turned head to wind away from the rig it rolled right over on its starboard side. Lionel had only time to release the inflatable life-rafts outside the wheelhouse before he was flung off the bridge into the sea, and as the boat started to sink he found he was trapped under the funnel. The water closed over his head and he was sure that when the boat went down it would take him with it but, as it rolled over, a surge of water washed him away and he found himself floating twenty or thirty feet away from the hull. He could see his crew and the men from the Hewett sitting on the side of the hull but, numbed with cold, he was unable to swim towards them. The sea was still very rough, but in the distance he could see a small trawler heading towards the *Hector Gannet,* and he prayed that someone would spot his bright orange life-jacket. As it approached, he heard someone shout and saw a figure point in his direction and within minutes he was being heaved aboard and wrapped in blankets.

The *Hector Gannet* sank rapidly, and the crew of the trawler picked sixteen men out of the sea and set off for Yarmouth. Though dazed and shocked, Lionel was worried about his crew, and when a list of the survivors was checked, Harry Goodison, the mate, Tommy Grimes, a deckhand, and Len Dawson, the second engineer, were missing. While the trawler was steaming for Yarmouth, the Arpet platform reported what appeared to be an orange life-jacket drifting close by.

'I've got a terrible feeling that's one of my lads,' said Lionel to the skipper of the trawler. 'Could we turn back? You never know!'

The skipper spun the wheel, and as they steamed towards the Arpet they spotted a body in the water and heaved it aboard. It was Harry. The crew carried him to the galley, and Lionel tried to revive him with mouth-to-mouth resuscitation. A doctor was lowered from a helicopter, but it was all in vain. Harry was dead.

When the trawler reached Yarmouth, the Reverend George Shelley, the Port Chaplain, and his wife Mary were waiting at the Seamen's Mission with hot drinks and dry clothes for Lionel and his men, but there was more sad news. Tommy Grimes had been lifted out of the sea by an RAF helicopter but was found to be dead. Len Dawson, the second engineer, was still missing and presumed to have been trapped in the engine-room when the *Hector Gannet* went down. Lionel was deeply depressed. It was a cruel twist of fate that three of his crew, all experienced seamen, should be lost and the men he rescued, including one with a broken leg, survive.

On the morning of Friday 15 November 1968, Flight Lieutenant Frank Davies was the duty pilot of 'D' Flight, 202 Squadron at RAF Coltishall, a few miles north of Norwich. 'D' Flight was a Search and Rescue Unit using Whirlwind helicopters, and though November had been a quiet month for the unit, with only a forced landing of a jet aircraft near Norwich on the 8th and an aircraft crash near a United States Air Force base in the south of the county on the 12th, the crews were always on full alert.

At 11a.m. the coastguard at Gorleston, near Yarmouth, telephoned with the information that a gas rig, fifteen miles east of Cromer Head, had 'blown its top' and that the supply ship alongside had capsized while taking off survivors. Within four minutes Frank Davies was airborne in Whirlwind Xray Kilo 969, with Flight Lieutenant John Pottage navigating and Flight Sergeant Gordon Ray as winchman. It was a bumpy flight in a gale-force wind, and for a long way they could see the blow-out spurting like a black fountain from the rig.

Approaching the Hewett, Frank Davies could see a trawler picking up objects in the water. A large Bristow helicopter and one from the British European Airways base at Lowestoft were already on the scene, hovering over a man lying face down in the water. All helicopters operating in the North Sea were fitted with floats to enable them to land on water if necessary, but the sea was too rough, and without winching-gear the pilots could do no more than hover above the man to mark his position until the RAF arrived. Frank Davies positioned the Whirlwind while Gordon Ray, the winchman, hoisted the man out of the water. He was wearing only slacks and a white shirt and, despite the efforts of Gordon Ray and John Pottage to resuscitate him during the flight to Yarmouth Hospital, was dead on arrival.

The Whirlwind returned to the scene with a doctor on board, and he was lowered onto the trawler *Boston Hornet* to examine another man who had been picked up out of the water, but he too was dead. The doctor decided that three of the survivors on the boat needed medical treatment and should be lifted off and taken to the Arpet. Communication difficulties were enormous. The trawler was not fitted with a radio that enabled the skipper to contact the helicopter, nor had the helicopter pilot any means of speaking directly to the skipper. Messages were conveyed by hand signals or relayed via the Arpet platform. The skipper was instructed to steam slowly downwind to reduce the motion of the trawler, but for Frank Davies manoeuvring the helicopter was more complicated. To remain airborne, it had to face the wind, which meant that, in order to keep pace with the trawler, the pilot had to fly the helicopter backwards.

The three men were successfully lifted off and landed on the Arpet, and Frank Davies returned to the *Boston Hornet* for the doctor. Ships reported two large life-rafts drifting near the Hewett, and the winchman was lowered to see if there were any survivors in them. They were empty, and to save any other false alarms, they were burst with a knife and sunk. All the crew off the rig and the *Hector Gannet* having been accounted for, with the exception of the second engineer, who was presumed to have gone down with the ship, there was nothing further for the RAF helicopter to do, and it returned to Coltishall. In his log the pilot wrote: 'This very ugly incident could well have resulted in a large loss of life, especially considering the cold strong winds in the area. Fortunately Hewett "A" is only some sixteen nautical miles from the Norfolk coast, but it is nonetheless a tribute to the rescue units that so many were able to survive.'

Friday 15 November was the Padre's day off. It had been a harrowing week at the Seamen's Mission in Yarmouth, and the Reverend George Shelley, the Port Chaplain, had promised his wife Mary that he would not answer the telephone, so when it rang at half past two in the afternoon, he ignored it. The ringing continued, and he burrowed deep into his armchair, hoping that whoever it was at the other end would go away. He knew that if he picked it up it was goodbye to his day off. He would have to put his book down, leave his warm fireside and go down to the Mission to talk for hours to a distraught fisherman's wife whose husband was drinking every penny he earned, or it could be any one of a thousand things he had to deal with every day of his life, except this one. He had not had a day off in weeks, and no one, just no one, was going to ruin it!

The telephone continued to ring. The Padre closed his book with a sigh of exasperation. How inconsiderate some people were! Perhaps he should read in his study, where there was no telephone. But it was chilly in there without a fire, and in any case why should he have to hide? It was his day off! It was written in the Scriptures for anyone who cared to read it: 'Six days shalt thou labour and do all thy work.' The ringing persisted and seemed to get louder. He looked at his wife, but she concentrated hard on her knitting. Years of experience had taught her that when the telephone rang it was never an invitation to a party. Some-one, somewhere, needed help. The Padre knew it too. He picked up the telephone.

'Hello, George Shelley speaking,' he said into the mouthpiece.

'This is Offshore Marine, Padre. Sorry to trouble you, but one of our ships has been sunk. One man is dead, and two are missing. The survivors are coming into port on the Lowestoft trawler *Boston Hornet*. ETA 1500 hours. Can you take over?'

It was a request the Port Chaplain knew only too well and had come to dread. 'Yes, of course, Captain,' he said. 'Can you give me any more details?'

'Yes. I'm sorry to say it's the *Hector Gannet*. She was on standby at the Hewett "A", and the rig had a blow-out. We're not sure yet exactly what happened, but it seems the *Gannet* was holed while the skipper was trying to evacuate the rig's crew.'

At the Mission it was the usual drill. Towels and blankets to be warmed on the radiators, spare clothes laid out and water-heaters switched on to make sure there was hot water for showers and baths. The Padre's wife prepared vegetables for home-made soup and switched on the tea urn. She telephoned a local girls' school where she taught, and the headmistress sent over some of the older girls to help out. Loading extra blankets into his van, the Padre drove to the harbour and arrived to find the *Boston Hornet* already tied up at the quay. Staggering under the weight of an armful of blankets, he pushed his way up the gangplank and onto the trawler, but before he had gone more than a few steps, he slipped on the wet deck and fell heavily onto his back.

'I hope you've got enough for sixteen of 'em,' said a crew member, helping the Padre to his feet and picking up the fallen blankets.

'Sixteen!' exclaimed the Padre in surprise. 'I was told it was only the crew of the *Hector Gannet*.'

'Oh, aye, we've got them,' said the crewman, 'and we've got eleven off the rig as well.'

As they were talking, a doctor who had been on board to check the men came out of the cabin. 'Padre, have you got anywhere where I can examine these chaps?' he asked. 'There's hardly room to move down there!'

'No problem,' replied the Padre. 'Bring them up to the Mission. You can use my office.'

The men were taken in ambulances and after a medical examination were given dry clothes and food. Many were in a state of shock, and one man could talk of nothing but losing his cap. He searched and searched through the supply of spare clothes and would not rest until someone eventually went out and borrowed one for him.

When the press and television crews discovered where the survivors had been taken, they descended on the Mission in hordes, bombarding the men with questions until they were so confused that they hardly knew what they were saying. The last survivor was brought in, and it was the skipper of the *Hector Gannet*. He was wet and cold and could not understand how a pyjama jacket he had been wearing had disappeared, while the sweater that covered it was still there. What he longed for was a hot drink and a shower, but the reporters pounced on him and harassed him with questions until he was so exhausted that the Marine Superintendent of Offshore Marine had to intervene and get him away. Not until the last member of the *Hector Gannet*'s crew had been attended to and fed, clothed and taken to a good hotel for the night did the Offshore Marine representative leave, and the owner of the rig's catering company showed the same concern for his men. Three of his young staff had scrambled down the nets onto the *Hector Gannet,* and they were utterly worn out after their ordeal. The catering boss whisked them away immediately to his own hotel and looked after them as if they were his own family.

No one came for the rig men.

The Padre and his wife provided hot showers, gave them food and clothes, let them use their telephone to ring their homes and generally kept them happy; then, after several hours of waiting, a

curt telephone message instructed the Padre to tell them to make
their own way to a hotel on the far side of the town. It was one
of the cheapest and dingiest places imaginable.

Some weeks later the head of the catering firm donated several mattresses to the Mission by way of gratitude for their
helping his men, and Offshore Marine sent a sizeable cheque.

Through the radio-room window I had watched, mesmerized
with horror, as the *Hector Gannet* steamed away, listing heavily.
Twice the boat had collided hard with the rig's legs, and the rig
swayed so violently I felt that any minute it would collapse into
the sea. As the boat turned, it rolled onto its starboard side, and I
could see a line of men sitting on the hull. At that moment
Vernon burst into the radio-room again.

'Bob, send out a Mayday. We've gotta get off this mother
afore she goes to pieces.'

I switched the transmitter to emergency battery power,
rammed home the red button fitted on all ships' radio sets which
transmits an automatic alarm signal on the maritime distress
frequency, waited a few seconds, then spoke into the microphone.

'Mayday. . . Mayday. . . Mayday. . . This is the oil rig Hewett
"A". Our standby boat *Hector Gannet* has capsized in position
52-59-58 north, 01.50.51 east. She is lying on her starboard side.
We ourselves are in trouble. Have gas blowout and we are preparing to abandon the rig.'

I released the transmit button, and there was an immediate response from the duty officer at the coastguard station at Yarmouth: 'Hello, Hewett "A". This is Gorleston coast-guard. What
assistance do you need?'

I could hardly hear his voice above the roaring noise outside
and the screech of the alarm sirens, and I clamped my headphones tightly against my ears.

'Hello, Gorleston,' I replied. 'There are men on the *Hector
Gannet* who need immediate assistance, and there are forty men
on the rig. We need to evacuate immediately. Can you alert
Bristow helicopters?'

'Roger Wilco!' was the brisk answer.

The Bristow base at Yarmouth had, in fact, heard my distress call, and helicopters were already on their way. During the few seconds the automatic alarm was being transmitted I had switched on the helicopter frequency radio and now gave out the Mayday with a microphone in each hand.

Humber Radio called and asked for the latest information on the *Hector Gannet*. As I rushed to the window, I saw the hull sink beneath the waves, leaving the men floundering in the water. I picked up the microphone.

'Hello, Humber Radio. This is the Hewett "A". The *Hector Gannet* has sunk and there are men in the water. Repeat. The *Hector Gannet* has sunk and there are men in the water.'

The trawler *Boston Hornet*, the Arpet platform's standby boat, was racing to the scene, and through binoculars I watched the crew pull a solitary figure in an orange life-jacket out of the water before picking up the rest of the men. In the rough seas and gale-force winds it was a magnificent display of courage and seamanship.

The rig was vibrating as if a giant hand was shaking it from side to side, and the noise was so intense that I kept my headphones on to protect my hearing. There was a strong smell of gas, and every time I pressed the transmit button on the radio I winced at the prospect of the wire antenna sparking. Humber Radio repeated my Mayday call, and it was reassuring to hear ships as far as thirty miles away reporting that they were steaming to assist. When I looked out, we appeared to be surrounded by ships of all sizes, but they could do no more than watch and wait! It was far too wild to attempt to come alongside. Helicopters arrived and were landing and taking off, and a yellow RAF helicopter winched one figure out of the sea and flew away.

I was busy giving Humber Radio an update on the situation when Vernon ran into the room.

'C'mon, Bob. We've gotta get the hell outa here!' he yelled, pulling me out of my chair. 'She's gonna go to bits any minute.'

'Humber Radio,' I shouted into the microphone, 'we're abandoning the rig and closing down.'

Flinging my headphones on the desk, I snatched up my windproof anorak and ran for the door. Out on deck it was a terrifying sight. A thick black column of mud, gas and oil was being blasted up the derrick and spewing drill pipe high into the air like pieces of straw. The noise was incredible! Deck, pipe stacks and companionways were plastered in an evil-smelling black mess that was being hurled about in the wind, and debris lay everywhere.

The full force of the wind almost blew us into the sea as we ran up the steps to the helideck, and when Vernon reached the top and hung on to the guard rail, I thought he had lost his balance. Gasping for breath, I pulled myself up to join him and, following his gaze, realized why he had stopped - the last helicopter had gone. The rig was deserted. We had been left behind!

As we scrambled back to the deck, a tangled mass of steel crashed only yards away from us, and we dived for cover.

'We'll launch a life-raft,' bellowed Vernon in my ear. 'One of the boats will pick us up.'

Keeping a wary eye on the derrick, we made our way to the west side, where the crew had dropped a landing-net over, but the cradles holding the inflatable life-rafts were empty. Vernon pointed across the deck. 'We'll try the east side,' he shouted. Chunks of debris were raining down on the helideck and on the pipe deck, and to avoid being hit we went through the accommodation and the galley. It took our combined strength to open one of the heavy steel doors on the east side against the force of the gale, and out on deck it was almost impossible to stand upright.

There were several life-rafts at this side, but looking at the enormous waves crashing against the side of the rig, we both knew it was a hopeless situation. Even if we could have launched a life-raft without its being blown away by the wind, it would certainly have been battered to pulp by the surf, and us with it. The rig was shaking violently and stacks of pipe were breaking loose and rolling on the deck. We crouched in the shelter of a deck cabin, and it seemed that at any minute the structure would collapse into the sea and we would die. It was just as my old sea-captain friend had said: when your number was up, there was nothing you could do about it. All around the rig were ships waiting to pick up survivors, but without some means of reaching them we might as well have been a hundred miles away. Unless we were picked up very quickly, jumping into the gale-lashed sea would have meant a lingering death from exposure.

Oddly enough, I did not feel the least bit frightened. I had a feeling of being like a ball on a pin-table. Life had rolled me around from one situation to another; sometimes I had landed in tricky situations and had managed to bounce out of them, but now I had the biggest score on the board, marked 'finish'. The strangest thoughts flashed through my mind - I was angry with myself for going fell-walking on my last shore leave and not mending my youngest daughter's doll's house as I had promised. I was angry that I had selfishly bought a boat when the carpets at home were threadbare and the children needed new clothes. I wanted to tell my aged and long-suffering parents that I was sorry for having caused them a lot of worry and heartache, and I wished I had taken their advice and got a secure job in an office.

In my youth I had been brought up in a strict Methodist home; I had rebelled against it and ridiculed it, and now perhaps I was about to pay the price. The words of a prayer formed in my head, but I fought them off. 'Damned hypocrite,' I told myself. 'You're like those people who sing hymns when a ship's going down - most of them would never normally go through a church door.' I was shaken out of my thoughts by Vernon's pulling at my arm and indicating that we should go back inside the accommodation.

The whole place reeked of gas, but it was a relief to be away from the noise and the biting wind and spray. Vernon was sure that, when it was discovered we were missing, a helicopter would be sent back for us, but he was worried that the derrick might fall and wreck the helideck. In the radio-room the helicopter radio and the emergency transmitter were still on, and Humber Radio was co-ordinating the search for survivors. The *Boston Hornet* had picked up a body near the Arpet platform, and the skipper asked for a doctor to be lowered on board from a helicopter.

It was very tempting to call the radio station and let them know we were still on the Hewett, but Vernon said that it only needed a spark from the antenna to blow us to kingdom come. There was a good deal of chatter on the helicopter frequency between the Arpet platform and an RAF rescue helicopter, and we were listening to it when a pilot's voice broke through, calling the rig.

'Hello, Hewett "A". I realize you may not be able to transmit; I will be at the rig in three minutes but I cannot land. I will hover as close to the helideck as possible, and you must get aboard as fast as you can.' He started to repeat his message - but we were already running for the door.

On the helideck the wind made it difficult to do anything except hold on to the guard rails, and at first we could see nothing but masses of spray thrown up from the sea. A few miles away the Arpet platform was ablaze with lights, and suddenly one appeared to detach itself and head straight for us. It was a large helicopter. Keeping well away from the pall of gas and mud over the rig, the pilot hovered a few feet away from the edge of the helideck. Between us and safety was a yawning gap with a long drop into the sea. The deck was slippery with spray, my legs and fingers were numb with cold, and wind was tearing at my clothes. I would never make it. 'Jump!' bellowed a man standing in the aircraft doorway. Taking a deep breath, I ran and made the leap of my life. Strong hands grabbed my anorak, and as I was bundled onto the floor, Vernon was hauled to safety behind me. The helicopter banked away, and through the window I caught a glimpse of an RNLI lifeboat wallowing in the sea below. We were safe, but for some the search was still going on.

At Yarmouth heliport it was utter confusion. Press and television crews descended like a swarm of locusts, and when the helicopter from the Hewett landed, the men were besieged by reporters and photographers, clamouring for a story. The injured derrickman and several others, suffering from exposure or the effects of the gas, were lifted into waiting ambulances, and the drivers had to bulldoze the vehicles through the frenzied pressmen before they were able to speed to the hospital.

The rescued crew had nothing with them other than the few clothes they were wearing, but there was no equivalent of the friendly Seamen's Mission waiting to care for them with hot drinks and food. A minibus was laid on to ferry them to the town, and they were dumped on the doorsteps of the cheapest hotels the company could find.

When Vernon and I landed at the heliport some time later, the pressmen had gone, and we took a taxi to the Rimrock office near the harbour. Vernon was ushered into an inner office to make his report, and I asked if I could phone my wife to tell her I was safe. 'Yeah, OK,' said the office accountant reluctantly, 'but make sure it's a collect call!' At first I thought it was his idea of a joke to break the tense atmosphere in the office, but he was deadly serious. I made my call and, blazing with anger at his stinginess, went out into the street. Connie, the office secretary, had given me the name of a hotel where rooms had been booked for the Hewett men, and I walked round the town looking for it.

I was eventually directed to a tatty and neglected building by the harbour and went in to find the men happily swigging pints of beer and large glasses of whisky, gin and vodka as if they had gatecrashed a pools-winner's party. Vernon, with his usual concern for the welfare of his men, had phoned the hotel to say that drinks were on the company. I ordered a very large whisky and wished I could have been at the side of the Scrooge-like figure of the office accountant when he received the bill for twenty or more rig workers on a drinking spree.

I was shown to my room by the hotel proprietor, a scruffy, unshaven, booze-sodden individual who cheerfully bragged that the Hewett blowout was the best thing that could have happened to his business, and that he was going to make a lot of money out of it. I had a mental picture of the *Hector Gannet* lying on its side, and I could happily have put my boot into his repulsive gut. The room was filthy, the washbasin was cracked and leaking, there was no hot water, and the sheets had not been changed after the last occupant. I went out onto the landing to find a toilet and met Louis Sylvester, who was equally unhappy about his room.

'Well, I'm not staying in this dump!' I said. 'Why don't we find out where the Phillips Petroleum people are staying and we'll book in there?'

'Good idea!' said Louis, and we walked into the town. Passing a phone booth, I called Phillips and asked where their top executives were staying. The receptionist told me the name of the hotel. It was the best and most expensive in town. 'That'll do for us,' I said to Louis, and we booked a room each, with *en suite* bath. We bought new slacks and shirts so that we would at least be reasonably presentable, and at dinner that evening the Phillips managers, who knew us, almost choked on their steaks when they saw Louis and me at the next table. It made our steaks taste all the better, and we washed them down with a bottle of the best house wine.

A certain name and telephone number are permanent fixtures on the deskpad of every oil-company executive involved in drilling. The companies may enjoy flexing their muscles, and hire and fire to show how tough they are; they may influence governments, screw the last penny out of a contract, drop their public relations company, change to a different helicopter operator or supply-boat company. They wield the big stick over the lives of hundreds of people, but if a company is in serious trouble with a fire or a blowout on a rig, there is only one man on this earth who can help them: Paul 'Red' Adair, from Houston in Texas.

Red Adair's reputation as a tamer of oil-rig fires out of control and gas blow-outs that threaten to annihilate the population, has made him a folk hero. Scores of magazine and newspaper articles have been written about his exploits, and no less a hero figure than John Wayne played the leading role in a Hollywood film about his life-story. Now Phillips sent Adair an urgent message and, like any other king, he arrived in style with his entourage and was booked into our hotel.

In the hotel a man who was a clever artist had painted a large invitation card in beautiful copperplate script, and hung it on the wall. It read:


Phillips Petroleum Company UK
request the pleasure of the company of
Mr Paul 'Red' Adair
on the occasion of their blowout on the Hewett platform
in the North Sea.

RSVP (Bloody quick!)

It was removed very rapidly by the hotel management.

Phillips executives, Red Adair and Rimrock discussed tactics, and it was decided that Red and Vernon would fly out and attempt to land on the rig. They needed someone to operate a portable radio to maintain contact with the helicopter while they were making an inspection. Saying 'no' to someone as friendly and persuasive as Vernon was never easy, and before I realized what I had let myself in for, a taxi was taking us both to the Phillips office and warehouse on the quay.

Having heard all about Red Adair, I expected to see a replica of Big Tex barge through the door, shouting his mouth off and playing at being a hero, but nothing could have been further from the truth. A thick-set man of average height with red hair and freckles and a broad smile on his face came into the room, and after shaking hands and introducing himself led the way to an adjoining warehouse. Lying on the floor were several suits similar to helicopter survival suits, portable radio sets, electric torches, divers' air bottles and toolboxes.

'OK, you guys,' said Red. 'I'm figurin' on going out to the rig in the mornin'. We'll land on the Arpet platform first to get an idea what shape the rig's in, and if it looks OK we'll go for it.' He picked up one of the suits. 'This heah's a special fireproof exposure suit, and I want ya to pick one that has a hood and face mask and fits real good. If things go wrong, it'll stand a helluva heat, and ya can jump in the sea wearing one of these and it'll keep ya afloat. Until we can figure out how much gas is lyin' about, we'll be breathin' air from the bottles.'

Putting the suit down carefully on the floor, he picked up a torch and pulled an adjustable wrench out of the toolbox. 'This heah lamp is spark-proof, so you don't need to worry about using it if there's gas about. And the same goes for the wrench. All the wrenches we're taking out are made from bronze alloy, so they won't make a spark if they're dropped onto steel.' He placed the torch and wrench on the floor. 'Any questions?'

'Yes,' I said. 'What will I be doing on the rig?'

He grinned, and picked up one of the portable radio sets. 'What you'll be doin' is usin' this,' he said. 'When the chopper drops us, he'll high-tail it away and fly around till we holler. Now I don't know what shape the rig is in, and we might have to get the hell off the mother pretty damn quick, so I wantya in contact with the whirlybird, OK?'

'OK,' I replied, and took the radio off him to check the controls.

Early the following morning the equipment was loaded into a van, and Vernon and I were driven to the heliport. The posse of newshawks was already waiting for us in force, but we were spared the ordeal of their questioning by ducking down in the van, and before they had twigged what was happening, a Phillips car arrived with Red Adair and we were airborne and on our way. I had looked at the Arpet platform almost every day from the Hewett and was surprised to find that, though operated by different companies, they were almost identical in design. A steward brought cups of coffee, and while we drank it Red examined the Hewett through binoculars.

There was still a massive black column spurting out at the top of the derrick, but the wind had remained in the south-east and was carrying it away from the rig. Heavy surf pounded the legs, and I could see the abandoned scrambling-net flapping aimlessly against the side. A short distance to the west of the rig an orange buoy marked the grave of the *Hector Gannet.* The whole place had an air of tragedy and foreboding, and I glanced at Red Adair, hoping he would decide that it was too dangerous to risk landing on the Hewett, and we would return to Yarmouth.

Red stared through his binoculars. 'What's the latest on the weather?' he asked one of the Arpet men.

'Well, at six o'clock they were giving south-easterly force six to seven with occasional rain showers,' replied the man.

Red rammed the binoculars into a case. 'OK, guys,' he said briskly, 'let's go for it!'

I gulped hard and followed him inside to put on my survival suit. An intercom had been fitted in the helicopter to enable us to talk to the pilot, and Red requested him to fly backwards and forwards close to the south side of the Hewett while he made sure that it was safe for us to land. Even in the helicopter there was an aroma of gas, and I felt sure Red would abandon the mission and tell the pilot to fly back to the Arpet - but Red Adair's reputation was not built on failure. 'OK, buddy,' he called to the pilot, 'put it down.' The Whirlwind banked towards the helideck, and I could hear the roar of the gas above the helicopter's engine. 'This crazy sod'll kill us!' I muttered to myself, but there was no turning back.

On the Arpet it had been agreed with the pilot that we would have to jump out while he hovered a few feet above the deck. Vernon heaved the door open, we lowered the toolbox on a rope and, pulling my face mask down and turning the air bottles on, I clutched the portable radio like a parachute and jumped to the deck. The second we were all out of the helicopter, it shot away as if it had been catapulted, and we were on our own. I felt as if we had landed on the moon. Around us the sea was white and deserted, and in the middle of it was the rig, roaring and vibrating like a volcano about to erupt.

In our discussion at the Phillips base, Red had outlined what he hoped to do if we landed. It was to be only a preliminary investigation, but there might be pockets of poisonous gas lurking around, so masks and air bottles had to be kept on at all times. While Red and Vernon went to the pump-room, I was to go through the accommodation with a gas-detector to see if there had been any dangerous build-ups. Treading warily, we made our way down to the deck on the windward side of the rig, and with hand signals Red indicated that he and Vernon were going to the pump-room and I should contact the helicopter. I was wearing headphones and a throat microphone to enable me to operate the radio whilst wearing the survival suit, and pulling out the antenna on the set I called the helicopter. The pilot responded instantly, and I reported that the two were going to the mud-room and I was heading into the accommodation.

Leaving the radio on deck, I took my torch and, hauling the galley door open went inside. It was an uncanny feeling. Everything lay just as it was before the blow-out occurred. Cups half full of coffee stood on the mess-room tables, a tray full of bacon on the hotplate, and another with fried eggs. Pans of potatoes and vegetables for the midday meal were on the cooker, sweets and cakes were half prepared on the stainless steel worktops; and a huge piece of beef for Archie's *Entrecôte chasseur* was lying on a wooden chopping-block. Urns were full of tea, coffee and orange juice. The newly arrived steward's bucket, mop and scrubbing-brush lay on the floor in the passageway. It was as if everything was in a state of suspended animation and at the touch of a button would come to life, and the room fill with people drinking coffee or going about their work.

I worked my way along the passageways, checking each cabin. There was no power on, and in the men's accommodation it was pitch black. The torchlight cast eerie shadows along the passageways, and with the dull roar of the escaping gas and the vibrating of the floor and walls it felt as though I was descending into the gut of a gigantic sea monster. I wanted to turn and run back into the daylight.

All the doors of the men's cabins were wide open, and bedding and clothes were strewn on the floor, hurriedly abandoned in the rush to escape from the rig. Holdalls and suitcases belonging to the men who had arrived on the crew-change helicopter lay unopened on bunks, and the illuminated dial of a small portable radio gleamed in the dark on the top of a locker. On every wall my torch shone on pictures of stunning girls taunting me with voluptuous breasts, arched bottoms and mystic caverns, but they seemed to insult the grave of the *Hector Gannet* only a short distance away and I closed the cabin doors on their obscenity.

It was stuffy inside the survival suit, and I longed to wipe the sweat off my face, but I dare not remove the hooded mask. Although the needle on the gas-detector had remained firmly on zero in every room I checked, I did not trust it and constantly pressed the test button to make sure it was working. Returning up the stairs, I went to the radio-room, and for the first time the gas-detector sprang to life. The amount registered was below the danger mark on the instrument, but as a precaution I opened the outside door and the portholes to let air circulate round the room. The emergency radio was still on, and though the battery-state indicator was showing 'low', Humber Radio was booming out of the speaker. My immediate impulse was to switch it off but, reluctant to touch anything electrical, I left the room and went to my cabin. Stuffing my wallet and cheque-book into a pocket in the survival suit, I clambered out on deck and collected the radio.

The helicopter was flying round close to the rig, but there was no sign of Red and Vernon. Since the gas-detector had not shown any dangerous accumulations inside the accommodation, and outside it did not so much as flicker at all, I reasoned that it would be safe to turn off my air bottles and remove my hood. The biting wind took my breath away and chilled the sweat running down my face, but it was a joy to breathe sea air again. It felt as if I had been on the rig for hours, and I had no idea what time it was; my watch had been left behind with a piece of paper on which I had been requested to write my Last Will and Testament and a note to my wife. To shelter from the wind, I shifted down to the galley and set up the radio inside the door.

The Whirlwind helicopters had a limited flying-time, and at the Bristow base in Yarmouth, Dick Drake, the controller, and Gerry Freeze, a Phillips manager, had the onerous task of ensuring that no matter how long we were on the rig a helicopter would be there to snatch us off if things went wrong. Time ticked by and a second helicopter arrived to relieve the one which had brought us out to the rig, but there was no sign of Red or Vernon.

To make myself useful, I went round checking that all windows and access doors to the deck were closed in case the wind changed direction and filled the accommodation with gas. It was awesome being near the huge column of gas blasting like a giant black waterspout 200 feet into the air, but I took comfort from the knowledge that if it did explode I would know little about it. Returning to my wind shelter inside the galley door, I switched on the radio and heard the Arpet calling me. Phillips had asked for a progress report but there was little I could tell them, other than that the gas was still belching out and that I had not seen Red or Vernon since they had left for the pump-room. When the third relief helicopter arrived and there was no sign of them anywhere on the rig, I began to feel concerned for their safety. Where they had descended to was a fearfully dangerous place, and anything could have happened.

Sitting alone on the rig, only a few yards away from a roaring bomb, and feeling thousands of tons of steel being shaken like a child's meccano set beneath me set my imagination racing. Maybe they had been overcome by gas! Could they have fallen into the sea? Had they been trapped under machinery? Were they lying there hoping I would come looking for them?

It was laughable that I should be worrying about a man who was a legend in the oil business and who had been taming rigs while I was still at school. But Vernon himself had said on many an occasion, 'In the drilling business, there's more heroes pushing daisies than pushing tools.' As this conversation went on in my head, I caught sight of a movement at the foot of the derrick, and to my intense relief the two emerged on deck. They had both removed their masks, which was a good sign that the wind was blowing the gas away from the rig.

'Howya doin', fella?' yelled Red, with a big grin on his mud-stained face.

'I'll feel better when I'm off this rig!' I yelled back.

'Yeah,' he shouted, 'you and me both. Bring the whirlybird in!'

I called the helicopter and told the pilot to collect us and that it was safe to land. We lifted the tools and the radio on board and climbed in after them.

'OK, buddy,' said Red into the intercom. 'Let's go for Yarmouth!'

'How long were we on the rig?' I shouted to Vernon above the noise of the engine. Twisting his arm, he showed me his watch. We had been on the Hewett for over four hours.

Landing at the heliport, it looked as if every newspaper reporter, press photographer and TV cameraman in the world had turned out to meet us. The moment the rotors stopped and we stepped out of the helicopter, they came at us like a stampede of buffalo, and it was more frightening than being on the rig! Vernon and I were elbowed out of the way so that the mob could get in closer to Red with cameras and microphones. It was complete bedlam. Dumping the tools and suits into the waiting van, we sat in the deserted passenger terminal with cups of coffee from the dispenser and watched the circus through the window.

'That's the price of fame, Vernon,' I said with a laugh.

'Yup,' replied Vernon quietly. 'Ah guess ah'm happy bein' a nobody.'

Red and Vernon were whisked away by Phillips, but before he left, Vernon gave me £10. 'Go and relax, Bob,' he said.

'I can't take this, Vernon,' I protested.

'Yeah, you take it. It's company money anyway, an' ah'll let 'em know ya damn well earned it.'

When I reached my hotel, there was a message for me to ring Rimrock. I felt drained and weary and needed a hot shower, but it sounded urgent. Lying on the bed, I picked up the phone and asked the receptionist to get the number for me. I was put through to one of the office managers. I thought he might be concerned that I had no money with me. I should have known better! He launched into a tirade about my squandering company money by moving into an expensive hotel; the hotel I had been booked in was good enough for the likes of me. Who in hell did I think I was, anyway? I let the phone drop to the floor, and sank back dog-tired against the pillow.

Life on the Hewett flashed through my mind. The fierce gales; the bitter cold; toolpushers who worked under enormous strain and went without sleep for days on end; the drillers and roughnecks so frozen they could hardly move at the end of a tour; the roustabouts drenched with spray and battered by gale-force winds as they struggled to repair fuel and cement lines. People like Charlie, fighting to stay alive; Pegleg's incredible courage; the tragedies of Graham and Larry, and the sadness of the mechanic whose son had been killed in Vietnam; the bravery of the helicopter pilots and the boat crews, and the bold fearless-ness of Red Adair. All these people I had been privileged to work with.

The office man babbling at the end of the telephone lived in a different world. A regulated, unexciting, uneventful but safe world, represented by columns of neat figures in a book. A rig was something stuck out in the sea where men made impossible demands on his budget and upset his calculations or disrupted his balances by working extra hours or making errors in paper-work. At the end of his working day, when the office closed at 5 p.m., he would go home to his fireside and watch television. The newsreel films would be full of gory conflicts in distant parts of the world or show unpleasant pictures of thousands of people starving to death. He might have seen the TV pictures of the Hewett tragedy that showed gas pouring out of the rig, the men being rescued and the distressing sight of a boat lying on its side or dead bodies being lifted out of a trawler in Yarmouth harbour. But all this drama was in someone else's world, far removed from his own, and a news film was just like any other drama movie - when it was over and the TV set switched off, it could be forgotten.

When I woke up, the babbling in the telephone handset had stopped and I replaced it on its rest. There was a portable radio on the bedside table, and I switched it on as an announcer was giving a time-check. It did not seem possible that I had been asleep for three hours. A lot of the news bulletin was devoted to the Hewett accident. There was a dramatic story about how Red Adair and two assistants had fought to control the blow-out but had been unsuccessful. If the journalists who elbowed us out of the way at the heliport had cared to ask, I could have told them that the only fighting I did on the rig was against the impulse to call the helicopter to take me off.

During the night the wind veered to the north and carried the escaping gas over the rig instead of away from it, so a return visit planned for the following day was considered too hazardous. Meanwhile the Hewett crew were having a whale of a time in Yarmouth. The Rimrock accountant had hastily put a stopper on the free drinks spree, but the TV pictures of the blow-out had touched the hearts of the townsfolk, and dancehall managers, café-owners and publicans generously let the men in without payment or provided a snack or a drink.

My two weeks tour ended before the weather improved sufficiently for another visit to the rig by Red Adair, and I went home to a rapturous welcome from my family. On the morning of the blowout my wife had been listening to the radio while making lunch, and had frozen with shock when she heard the newsreader say there had been a blowout on the North Sea drilling rig

Hewett "A" and that it was not known if there were any survivors. She immediately called the Rimrock office, only to be told by some insensitive idiot that if I had not survived she would be informed in due course.

When I returned to the rig, the drama was over, and for the first time I discovered what had caused the blowout. While the drilling crew were pulling the last few sections of pipe out of the hole, for some reason - which neither Rimrock nor Phillips would enlighten me on - the mud pressure in the well fell below the pressure of the gas, and the mud was blown out of the hole. The blowout-preventer was operated to close the gap between the drill pipe and the casing, but the gas continued to force its way through the holes in the drill bit and blasted out through the drill pipe being unscrewed section by section on the drill floor. During our first trip out to the rig Red Adair and Vernon had discovered that a few sections of pipe and the drill bit were still jammed in the hole. If the three holes in the drill bit could be plugged, the gas would be cut off and the blowout ended.

The day after I left for home, the wind had gone to the south again and Red Adair, together with Vernon, Ron Gower and the roustabout crew, was flown out to the rig. The mud pumps were started, and a quantity of small rubber balls were mixed with the mud and pumped into the casing. By a combination of pressure and Adair magic, the mud and rubber balls were forced down between the casing and the drill pipe and into the perforations in the drill bit, and the gas was shut off as if someone had closed a valve. The pumps were kept going to try to get as much mud as possible in the hole, but the well was not going to give up without a fight.

The following week the skeleton crew had to be hurriedly evacuated when the well erupted again and threatened another blowout. Red Adair and the crew returned the next day, and by the end of the week Phillips announced that the problems had been overcome and the well was dead. That incomparable rig-tamer Red Adair had triumphed again!

By December the drilling programme on the Hewett was nearing its end, and with it, Rimrock's contract and my job. When my pay cheque arrived, I was astounded to discover that the ten pounds Vernon had given me when we returned from the rig with Red Adair had been deducted from my pay. Another company contacted me with an offer of a job as radio operator, but I was thoroughly disillusioned with Rimrock's uncaring attitude and I felt that if they were typical of the industry then there was no future in it for me.

On my final day on the Hewett I went ashore by supply boat. I had always found flying in helicopters rapid, noisy and impersonal, and for my last journey from the rig I wanted to feel the movement of a boat and savour the atmosphere of the North Sea. As the *Smit Lloyd 6* was about to leave, there was a shout from the rig and Vernon came racing down to the boat landing and shook my hand.

'It's been a pleasure knowin' ya, Bob,' he said warmly.

'It's been a great pleasure knowing you as well, Vernon,' I responded. 'All the best!'

'See ya in Hell,' he called.

With a roar of engines the boat pulled away, and I looked back at the rig. The seas breaking against the legs and the mud-spattered accommodation block were a reminder that, in offshore drilling, disaster was never far away. This time the great pressure of gas that had threatened to blow the rig and everyone on it to oblivion had been subdued; but there would be others. All it needed was one mistake, a faulty valve, a careless spark, a tired driller or overworked roughneck or roustabout, and the airwaves would resound again with calls for help.

The forces of nature might be tamed but can never be beaten. An orange buoy floating over the resting-place of the *Hector Gannet* was testimony to that.

There were many calls for a public inquiry into the blowout and in a newspaper interview Eric Lubbock MP (now Lord Avebury) said, 'The oil and gas business, like fishing, may never be free from hazards, but there are some puzzling features in this episode that some MPs are determined to elucidate.' Despite the demands of the Members of Parliament, however, a formal investigation into the cause of the blowout was never carried out.

Newspapers were full of dramatic accounts and pictures of the blowout, and Red Adair's heroic attempts to tame it featured on television and in radio bulletins throughout the world. But there was one hero of the Hewett incident which the press, radio and television overlooked. When the Reverend George Shelley, Padre at the Seamen's Mission in Yarmouth, slipped on the wet deck of the *Boston Hornet* while hurrying to take blankets to the survivors, he hurt his back. Ignoring the considerable pain, he carried on throughout that terrible day caring for the wet, exhausted and distressed crewmen from the *Hector Gannet* and the rig. It was subsequently discovered that the fall had irreparably damaged his spine and he has been confined to a wheelchair ever since.

## POSTSCRIPT ...... 30 YEARS ON

After Red Adair saved the Hewett "A" in 1967, drilling started up again and in 1969 gas from the Hewett Field flowed through the 30" pipeline laid by the barge *Hugh W. Gordon* from the Hewett and Arpet platforms to the Phillips terminal 20 miles (32 kilometres) west of Hewett at Bacton on the Norfolk coast.

Gas production from these wells played a vital role in the conversion to natural gas in the UK, and in the 1970s contributed over 20% of the UK's gas supply. By the 1990s production had passed its peak and dropped to only 2%, but the field has had a considerably longer life than expected and the Hewett complex has increased to four production platforms, a field terminal and a 44-bed accommodation platform.

There are no visible reminders of the gale-swept November day in 1967 when the latest sophisticated offshore technology was beaten by pent-up gas a million years old and all hell was let loose, but not far from the Hewett "A", 100 feet (about 30 metres) below the surface, the ship *Hector Gannet* still lies on the sea bed, a monument to Harry Goodison, the Mate, Len Dawson the second engineer and Tommy Grimes a deckhand, the three brave seamen who lost their lives in the attempt to evacuate the crew from the rig.

By the early 1970s, considerable quantities of oil were being discovered off the Shetland Islands of Scotland and, as the drilling companies began concentrating their operations in this sector of the North Sea, Aberdeen, a large harbour and town on the east side of Scotland, replaced Great Yarmouth as the main base for UK offshore exploration and production. The prosperity bubble that had catapulted the name of the almost-forgotten fishing port of Great Yarmouth into the world's headlines and made many people rich had finally burst.

Having escaped unscathed from the Hewett "A", I vowed never to work on offshore rigs again and I fitted my boat out for lobster fishing and worked the lobster grounds along the Cumberland coast. But fate can play strange tricks, and a few months later when the engine failed the boat was blown onto a reef in heavy seas and wrecked. I was thrown into the sea and by a fluke of wind and tide managed to reach the shore. The boat was not insured and, utterly broke, I hitched a lift to Aberdeen and worked as radio operator/medic on semi-submersible rigs drilling the waters north east of Shetland. It was winter time and gales of force 9 and 10, often increasing to hurricane force, blew up with alarming frequency; and it was not unusual to have the rig covered in several inches of snow and the lattice work of the derrick encased in ice. For the drilling crews and roustabouts working on deck it was sheer misery, and when I had to make radio calls to the States and I talked to operators on Miami Radio who complained that they were having a heat wave, it was though I was in touch with another planet.

Scientists have claimed that, because the North Sea is relatively shallow, it cannot produce large waves; but had they witnessed the gigantic foam-capped mountains of water that day after day bore down on our floating metal island off Shetland during Christmas and New Year 1974, they would quickly have changed their minds. At the height of a storm that gusted to force 12 and above, I witnessed one mammoth wave bearing down on us that lifted the rig like a toy boat - at that time it was reputed to be the largest floating rig in the world - snapped 12 massive anchor chains as if they were cotton, buried the deck under hundreds of tons of water and carried away a 30ft steel portacabin, welded to the deck, as if it were a cardboard box. At its most violent, the power of the North Sea was awesome.

It was the most arduous time I had ever experienced, and when an opportunity arose to set up my own business repairing and maintaining boats at Ravenglass, my home on the Cumberland coast, I quit the rigs for good; but before I reached the shore the North Sea had a final sting in its tail to give me a send off I was not likely to forget.

The crew change helicopter on my final journey took off from the rig in a typical blizzard of snow, and as it approached Sumburgh airport in Shetland one of its engines suddenly lost power and it dropped like a stone towards the sea. By a superhuman effort the pilot managed to keep going on one engine and cleared the top of the cliffs with only feet to spare. We bounced along the runway flanked by a posse of fire engines and ambulances, and rolled to a stop shaken but unhurt. But that wasn't the end of the drama. The turbo prop aircraft taking us on the last leg of the journey developed engine trouble after take off, and was forced to descend and fly low over the sea all the way to Aberdeen, where there was another posse of fire engines and ambulances lined up in answer to the pilot's emergency call.

The search for oil in the Scottish sector of the North Sea was stepped up throughout the 1970s but, although there were significant improvements in technology and the UK Government directed that rigs should have qualified medical staff aboard, two incidents occurred which illustrated beyond doubt that it was still an industry in which the accident potential was very high.

In March 1980 one of the legs on the Norwegian platform, 'Alexander L. Keilland', located off Stavanger, Norway, collapsed, and the whole accommodation package fell into the sea with the loss of 123 lives. It was a dreadful tragedy, but in July 1988 worse was to come.

Occidental Petroleum Company's 'Piper Alpha' platform, 110 miles NE of Aberdeen, caught fire and was enveloped in flames so intense it melted steel into liquid and took the lives of 167 of the 255 men on board. It was described as the worst offshore rig disaster in history.

Following these tragedies, stringent guidelines for offshore drilling operators were introduced and safety regulations tightened, but though many hard lessons have been learnt since offshore exploration began in the North Sea over 30 years ago the one certainty about drilling in hostile waters is that wherever there are rigs there will be men and metal pushed to the limit. When men and metal are pushed to the limit something has to give.

## About the Author

Robert Orrell has had a lifelong connection with the sea, having previously worked on trawlers, as a lighthouse keeper, Merchant Navy Instructor, lobster fisherman and yacht skipper in addition to his job as radio operator on the Hewett "A". He has also been a BBC radio producer, made television documentaries and written several books on topics as diverse as adventurous pony trekking in the English Lake District, the Scottish Highlands and the Isle of Man, a children's book about the Lake District, guide books to places of interest in the Lake District and the history of the UK's largest Livestock Auction Company.

# INDEX